T0205500

# Determining Sample Size and Power in Research Studies

J. P. Verma
Priyam Verma

# Determining Sample Size and Power in Research Studies

A Manual for Researchers

 Springer

**J. P. Verma**
Sri Sri Aniruddhadeva Sports
University
Chabua, Dibrugarh, Assam, India

**Priyam Verma**
Department of Economics
University of Houston
Houston, TX, USA

ISBN 978-981-15-5206-9     ISBN 978-981-15-5204-5    (eBook)
https://doi.org/10.1007/978-981-15-5204-5

This Springer imprint is published by the registered company Springer Nature Singapore
Pte Ltd.
The registered company address is: 152 Beach Road, #21-01/04 Gateway East, Singapore
189721, Singapore

This book is dedicated to

The research scholars

# Foreword

It has been my pleasure to go through the contents of the book titled *Determining Sample Size and Power in Research Studies: A Manual for Researchers* written by J. P. Verma and Priyam Verma. The topic of the book is important for applied researchers as the authenticity of findings depends upon the sample size taken in the study. In studies with large sample, even small effect becomes significant, whereas in small sample a large effect may not be significant. This puts a researcher in a dilemma as to what should be the sample size. A notion which prevails amongst researchers that large sample gives reliable findings in comparison to that of small sample may be misleading.

In reality, size of the sample should be decided on the basis of how much effect is warranted for specified power and error rate in the test. The smaller the effect size, the larger the sample required, all else equal. Power of the test also increases with the increase in sample size, all else equal. But enhancing power increases Type I error rate. Due to the co-dependence of these three parameters, one can estimate the sample size required in the study for a given effect size conditional on the power and the error rate. Population variability which is obtained either from similar studies conducted earlier or by organizing a pilot study is also an important parameter which dictates the sample size for a desired effect. The larger is the variability, the higher is the sample required, and therefore researchers delimit their study to minimize population variability.

Multiple parameters and their interdependence could potentially be confusing for an applied researcher. In this respect, this book has done a wonderful job in providing practical solution to the applied researchers in estimating sample size in their study. In my opinion, this book will prove to be a milestone in providing guidelines and solutions to the researchers in enhancing validity in their research findings. The most important feature of this book is to provide solution to the researchers in determining sample size by using the freeware G*Power software. All 21 illustrations discussed in the book for determining sample size and power using different statistical techniques will help a large number of researchers in a variety of situations in different fields of the study.

The book has been organized logically into seven chapters. ► Chapter 1 provides a general description of the subject matter, whereas ► Chap. 2 explains various terms involved in conducting empirical studies. These two chapters together provide sound fundamentals to the readers for determining sample size and power in the experimental studies. ► Chapter 3 deals with survey analysis; hence, the field researchers would be benefitted in deciding the sample size for specific accuracy in estimating different parameters. ► Chapter 4 explains the theoretical concepts in determining the sample size and power in empirical studies. ▶ Chapter 5 explains the procedure in installing freeware software G*Power for determining sample size and reporting power in the results. Chapters 6 and 7 explain the procedure of determining sample size and power in varieties of applications. This makes the book very useful for the researchers in different fields.

The book satisfies the need for applied researchers and will be of immense help to:

- *Students*: As it will help them to determine sample size and power using G*Power software in their study.
- *Survey scientists*: To plan the size of sample for specific accuracy required in the study.
- *Reviewers*: To assess the validity of findings in research papers by checking the achieved power of the test for the given effect size.
- *Researchers*: For ensuring the testing of specific effect size in their empirical research.

The authors need to be complimented for producing such a valuable piece of work which can be used by the students, faculty and researchers in different areas of learning globally.

**Prof. D. P. Agrawal**
Ex-Chairman, Union Public Service Commission
New Delhi, India

# Preface

The philosophy of research is isomorphic to that of a sport. It comes along with its rules, violation of which leads to a heavy penalty. In this book, we will provide training instruments for researchers so they can be aware of the assumptions and conditions under which the statistical findings of their empirical work are valid. Statistical techniques have been widely used to provide weight to various hypotheses which are borne out of theories as well as experimentations. A virtuously conducted statistical test could form the foundation of knowledge base, which can be then used by policymakers, drug companies and future researchers.

However, as in a sport, if a rule is violated, the outcome of the study should not be recorded as a part of knowledge base. Even though great many findings have come about as a result of judicious statistical experiments, there exists a large body of empirical work which violates the assumptions of the statistical tests, and hence should not be used as a foundation for policies or future research. What the sample size should be in empirical studies remains the first step once the question is defined.

For survey studies, this becomes more important as the cost of including say 50 more observations could be quite high. Hence, knowing the cost feasibility of a study based on the sample size requirement could be very important. Based on statistical theory, there are three factors which determine the optimal size of the sample: The accuracy required in the estimation, power of the test and population variance. Small sample size may be unable to detect the effect, while the large sample could suggest a significant effect even when the effect is very small.

However, in most practical cases, small sample size concerns the researcher. In which case of the three factors listed above, power of the test remains of utmost importance. In lucid terms, it is the confidence with which the researcher can reject the null hypothesis. If the power of any test is 50%, which implies that the statistical test adds no value to the information that the researcher had before the test. A power of 80% is generally acceptable in most studies. Thus, when researchers are working with small sample size, it is important that they report the power of the test.

In this text, we have discussed the concepts involved in sample size determination in survey studies as well as in hypothesis experiments. The procedure has been shown by solving some of the solved examples manually. Simultaneously, we have shown the procedure of determining the sample size in different situations by using the G*Power software, which is a freeware and can be easily downloaded. The whole book consists of seven chapters.

▶ Chapter 1 is introductory that discusses the importance of sample size determination, whereas the ▶ Chap. 2 discusses the theory of inference. In this chapter, we have discussed the concept of estimation and hypothesis testing. The role of effect size in sample size determination and the procedure involved in hypothesis testing experiments by fixing different boundary conditions have been discussed in a crisp manner.

▶ Chapter 3 discusses the procedure of determining sample size for estimating population mean and proportion in survey studies. Several solved examples have been illustrated to describe the procedures. In ▶ Chap. 4, we have discussed the sample size determination procedure in hypothesis experiments. Solved examples in the case of one- and two-sample testing have been illustrated manually to describe the procedure.

▶ Chapter 5 explains the procedure for downloading the freeware G*Power software and its installation on the computer. In ▶ Chap. 6, we have discussed the procedure in determining sample size as well as the power in studies using one and two sample tests along with testing significance of correlation and association using G*Power software. ▶ Chapter 7 discusses the procedure of estimating sample size and power in studies using multiple regression, logistic regression, one and two way ANOVA, repeated measures ANOVA and MANOVA techniques. The procedure has been explained by using the screenshots which will be helpful for the researchers to apply even without any statistical and com-puter background.

We hope that this book will encourage the researchers to determine sample size for specific power in their studies, and they will start reporting the power in their studies for authenticity in the findings.

The readers are encouraged to send their suggestions and queries to J. P. Verma at vermajprakash@gmail.com or to Priyam Verma at priyamverma.2992@gmail.com, and we shall respond to them at the earliest.

**J. P. Verma**
Sri Sri Aniruddhadeva Sports University
Chabua, Dibrugarh
Assam, India

**Priyam Verma**
Ph.D. Scholar in Economics
University of Houston
Houston, TX, USA

# Acknowledgements

We would like to express our gratitude to our professional colleague Prof. Harinder Jaseja who has not only helped us in editing and checking the manuscript but also added his inputs through discussions on many topics. We wish to express our sincere thanks to Prof. Y. P. Gupta and Prof. V. Sekhar for providing their valuable inputs in completing this text. We are extremely thankful to Mr. Deependra Singh for his support in preparing the manuscript and completing other formalities regarding its publication. We are also thankful to Mr. Amey Bhojane for his creative support.

We wish to express our sincere thanks to all the research scholars and numerous researchers in India and abroad who had presented a variety of questions on sample size determination and power analysis which helped us in editing the contents of this book.

J. P. Verma
Priyam Verma

# Contents

# About the Authors

### Prof. J. P. Verma

is the founder Vice Chancellor of the Sri Sri Aniruddhadeva Sports University of Assam. This is a state university of Assam Government established at Chabua in Dibrugarh. The university is a high class university dedicated for the sports education and research activities in north eastern region of India. Prior to this assignment Professor Verma was Head, Department of Sport Psychology and Dean of Student Welfare at Lakshmibai National Institute of Physical Education, Gwalior. He has more than 38 years of teaching and research experience. He also worked as the Director of the Centre for Advanced Studies three years. He holds three master's degrees in Statistics, Psychology and Computer Application and a Ph.D. in Mathematics. Prof. Verma has published eleven books on research and statistics in the area of management, psychology, exercise science, health, sports and physical education, and 45 research papers/articles, and has developed sports statistics as an academic discipline. He was a visiting fellow at the University of Sydney in 2002 and has held academic visits in universities in Japan, Bulgaria, Qatar, Australia, Poland and Scotland, where he conducted numerous workshops on research methodology, research designs, multivariate analysis and data modeling in the area of management, social sciences, physical education, sports sciences, economics and health sciences.

### Priyam Verma

is currently pursuing his Ph.D. Economics at the University of Houston, Texas. He completed his M.Phil. in Development Economics and masters in Economics at Indira Gandhi Institute of Development Research (IGIDR), Mumbai. He works in the intersection of public finance and trade. His research aims to infer about government's behavior using econometric techniques thereby validating existing economic models.

# Introduction to Sample Size Determination

© Springer Nature Singapore Pte Ltd. 2020
J. P. Verma and P. Verma, *Determining Sample Size and Power in Research Studies*,
https://doi.org/10.1007/978-981-15-5204-5_1

**1**

- **Learning Objectives**

After going through this chapter, the readers should be able to

- Understand the importance of deciding sample size in research.
- Explain as to why neither small nor larger sample is appropriate for empirical research without deciding the effect size.
- Learn the parameters that are required for deciding sample size and computing power in research studies.

## Introduction

Research is generally conducted to examine characteristics of a population. Studying entire population becomes expensive hence we examine a sample drawn from the population of interest to draw inferences. Except in census studies, all research is carried out on the basis of a sample. Sample studies, if done carefully, are more efficient than the population studies as they are economical and consume less time. However, the essential pre-requisite for statistical efficiency of any study is the representativeness of the chosen sample.

Large samples closely approximate the population and thus provide better inferences about the population. Small sample is considered less reliable as it may be representative of only a part of the population. So how big a sample should be? This is a frequent question asked by the researchers. Many times, researchers are guided by an arbitrary thumb rule to take about 30 sample points, however, that may not be optimal in many situations. Optimal sample size depends upon many considerations such as statistical test used, precision of the measures and design of the study. In the later chapters we will provide precise formulas for calculating optimal sample sizes in various situations which takes into account the above considerations.

In a scientific research, sample size is determined in one of the early phases. Two investigations conducted using the same methodology and providing the same using different sample sizes may indicate different decisions. While going through the research articles, the readers should be concerned about how the sample size was calculated as it directly influences the findings. Tests with very small samples may not detect the difference, while tests with large samples will have enhanced power that translates even a small difference into statistical significance. Both these situations are of little practical value to the researchers. Thus, sample should neither be too small nor excessively large for a desired benchmark.

The larger the variability of the measures in population, the larger would be the required sample size. Besides variability, optimal sample size also depends upon how much accuracy we require in estimating the characteristics or parameters and how much confidence we wish to have in the estimation. In experimental studies, we want to know the power of a test to detect specific amount of treatment effect. Formally, this is going to be given a function of both, Type I and Type II errors. These issues have been explained in detail in ▶ Chaps. 3 and 4.

# Issues with Very Small Samples

The power of a study can be defined as ability of a test in rejecting null hypothesis correctly. In other words, if the effect exists then the test should be able to detect it. Small sample can detect a large effect but fail to detect a small effect. Suppose a study is conducted on patients to see whether a new drug enhances haemoglobin level more than that of the existing drug which the doctors normally use. Similar patients in terms of age gender and haemoglobin levels will be randomly assigned to two groups: one group will be given the existing drug and the other will be given the new drug. We know that for a specific power, 70 subjects (35 subjects in each group) are required for the study so that the findings of the sample study can be approximated to the population with sufficient power. The researcher finds that the new drug is more effective in increasing haemoglobin than the existing drug. Consider that the researcher instead uses 15 subjects in each group, and the study shows that the null hypothesis is retained. Let us investigate the implications. Using a sample smaller than the optimum increases the chances of rejecting the correct claim. As a result, power of the test decreases. Although the new drug is effective, it is not revealed due to the smaller sample size. The other drawback will be that the patients were unnecessarily exposed to hardship with inconclusive findings. Besides this, experimenter wasted financial and time resources, and the study will have to be repeated again for conclusive findings for a desired significant effect with larger sample.

# Issues with Very Large Samples

Usually, researchers believe that the large samples provide better results in research studies. In fact this is not generally true. Taking samples that exceed the number estimated by the sample size calculation poses different problems. The most vital issue is ethical. For instance, in medical studies, should we perform the study with the number of patients larger than required? This means exposing patients unnecessarily to the new drug resulting in increased risk. The situation becomes more serious if the new drug is inferior to the existing drug because more patients are involved with new drug therapy that yields inferior results. The second issue in using larger sample is that one needs to use more financial and human resources than required to obtain the desired response.

Besides these issues, there is another crucial issue that is related to the statistics in using the large sample. All statistical tests have been developed to deal with samples and not populations. In using the sample size larger than the required, the power of the test is substantially increased. This enhances the chances of rejecting null hypothesis for very small effect which may not be clinically useful. A hypothesis which is insignificant with the small sample size becomes significant now. Thus, a potential statistically significant difference in the effectiveness of new drug in comparison to the existing one may not warrant to treat the patients' with the new drug as the scale of superiority of the new drug over the existing drug may not be of any practical utility.

**1**

## Strategy in Sample Selection

If a large sample is available in retrospective studies, then select the subsample of the required size as determined by the requirements discussed in ▶ Chaps. 6 and 7 and thereafter perform statistical analysis, whereas in prospective studies one should select the sample that is required. If there are chances of heightened risk during an experiment, then few additional samples may be selected.

## Common Errors in Conducting Research

In undertaking a research study, researchers fix Type I error (chances of rejecting null hypothesis when it is true) and test the null hypothesis. In such a situation, if null hypothesis is rejected, one cannot be sure about the validity of research hypothesis. To explain this fact, let us consider an experiment in which an investigator is interested to know whether health awareness programme is effective in reducing obesity of participants in 4 weeks of duration. She selects a sample of 50 men and women who are in 100+ kg weight category. The initial weight of the subjects is recorded, and then a 4-week schedule is provided to them. After the experiment is over, their weight is again recorded. Applying the paired $t$-test at 0.05 level of significance, she rejects the null hypothesis of no difference between the pre- and post-average weights and thus claims that her 4-week programme is effective at 5% level. In other words, she is 95% confident that the health awareness programme is effective in reducing obesity. Another interpretation about her finding is that 95% of participants who join her programme reduce their weight; hence, people may consider her programme if they wish to reduce their weight. The question is should the prospective clients join her programme for reducing their weight. Here, there are four different issues with the findings of this study: Firstly, on an average how much weight was reduced; secondly, what is the power of the study; thirdly, whether this finding is applicable to all the weight category people; and finally, whether the reduction in weight is due only to the programme. In the absence of information on above-mentioned four issues, the finding of the study has little meaning.

The first issue is related to the magnitude of effectiveness (effect size), whereas the second issue indicates the probability of correctly rejecting the null hypothesis (power). Unless there is sufficient effect and power in the study, the result of the study is meaningless. The third issue is related to ethics of the study. Since subjects in the study were in 100+ kg weight category, the researcher should report that the finding is meant for such subjects only. On the other hand, the fourth issue is related to the internal validity of the findings. In order to ensure that the weight loss was due to the intervention only, the researcher should explain the design of the study.

In most of the studies, $p$-value is reported and if this value is lower than our choice of alpha ($\alpha$), we reject the null hypothesis. Hence, $p$-value indicates that the effect exists but the effect size indicates how much it exists. Further, if power in the experiment is not sufficient, there is no point of organizing the hypothesis testing experiment. For instance, if the power of the test is 50% or less, then it is better to toss a coin in rejecting or retaining the null hypothesis instead of conducting an experiment. As a thumb rule, the study should have at least 80% power. All these issues shall be further discussed in detail in the upcoming chapters in the book.

## Flow Diagrams for Deciding Sample Size

◘ Figures 1.1, 1.2 and 1.3 are the indicative flow diagrams in deciding sample size and achieved power in survey as well as experimental studies.

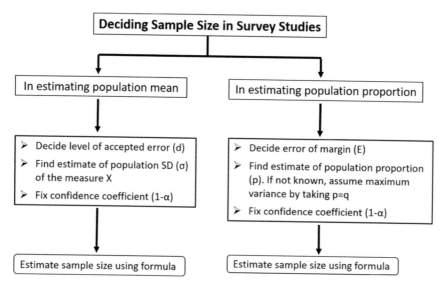

◘ **Fig. 1.1**  Flow diagram of computing sample size in estimating population mean in survey studies

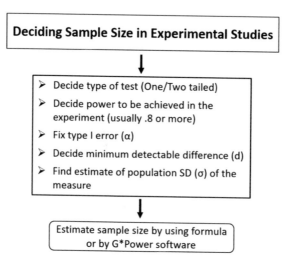

◘ **Fig. 1.2**  Flow diagram for estimating sample size in experimental studies

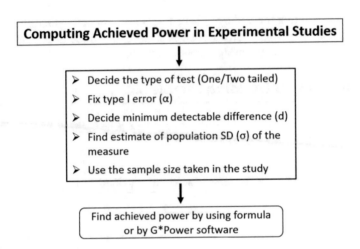

**◻ Fig. 1.3**   Flow diagram for estimating achieved power in experimental studies

## Summary

In this chapter, importance of sample size in conducting research has been discussed. Determining sample size in research studies is not only important for the statistical reason but also for the ethical as well as financial and human resource considerations. Issues with the small and large samples have been discussed using examples. Optimal sample size is determined either on the basis of precision or power in the study. The concept of precision is used in survey studies, whereas power is used in experimental studies. Finally, flow diagrams have been shown as a guideline to the readers in determining the sample size and power in research studies.

- **Exercises**
Q1. Why it is important to decide sample size in research studies?
Q2. Comment on the statement, "Large sample gives authentic results".
Q3. If the test statistic does not reject the null hypothesis for a given sample, should you increase the sample size and test the hypothesis again? Discuss the issue.
Q4. What are the problems with very small and very large samples in research studies?
Q5. What do you mean by delimitation and limitations in research studies? Do they affect sample size required for the study? If so how?
Q6. What are the usual mistakes which may be committed by the researchers in hypothesis testing experiments?

# Bibliography

Bartlett, J. E., II., Kotrlik, J. W., & Higgins, C. (2001). Organizational research: Determining appropriate sample size for survey research (PDF). *Information Technology, Learning, and Performance Journal, 19*(1), 43–50.

Cohen, J. (1988). *Statistical power analysis for the behavioral sciences* (2nd ed.). ISBN 0-8058-0283-5.

Faber, J., & Fonseca, L. M. (2014). How sample size influences research outcomes. *Dental Press Journal of Orthodontics, 19*(4), 27–29. ▶ https://doi.org/10.1590/2176-9451.19.4.027-029.ebo.

Fugard, A. J. B., & Potts, H. W. W. (2015). Supporting thinking on sample sizes for thematic analyses: A quantitative tool. *International Journal of Social Research Methodology.* ▶ https://doi.org/10.1080/13645579.2015.1005453.

Gambrill, W. (2006, June). False positives on newborns' disease tests worry parents. *Health Day.* 34471. html(dead link).

Ioannidis, J. P. A. (2005). Why most published research findings are false. *PLoS Medicine 2*(8), e124. ▶ https://doi.org/10.1371/journal.pmed.0020124.

Lachin, M. (1981). Introduction to sample size determination and power analysis for clinical trials. *Controlled Clinical Trials, 2,* 93–113.

Sterne, J. A., & Davey, S. G. (2001). Sifting the evidence—What's wrong with significance tests. *BMJ, 322,* 226–231.

Wacholder, S., Chanock, S., Garcia-Closas, M., Elghormli, L., & Rothman, N. (2004). Assessing the probability that a positive report is false: An approach for molecular epidemiology studies. *Journal of the National Cancer Institute, 96,* 434–442.

# Understanding Statistical Inference

© Springer Nature Singapore Pte Ltd. 2020
J. P. Verma and P. Verma, *Determining Sample Size and Power in Research Studies*,
https://doi.org/10.1007/978-981-15-5204-5_2

**2**

- **Learning Objectives**

After going through this chapter, the readers should be able to

- Understand why the distribution of sample mean becomes normal for large sample even if the population is not normally distributed.
- Identify the situation in using $t$- and $z$-tests.
- Learn the computation of confidence interval of mean and understand the factors that affect it.
- Interpret different terms in hypothesis testing such as Type I and Type II errors, level of significance and power of the test.
- Describe the procedure in testing of hypothesis by fixing effect size and power.
- Effect size and its use.

## Introduction

In inferential statistics, either we estimate population characteristics such as mean or some ratio using a representative sample or test their equality to some predefined values. Estimating parameters comes under theory of estimation, whereas testing parameters is covered under theory of hypothesis testing. One of the assumptions which is common to all the parametric tests is that the sample has been drawn from a normal population. One of the common parameters estimated in survey studies is the mean of the population. To rationalize the normality assumption in the parametric tests based on sample studies, let us see how the sample mean is distributed if a sample of the same size is repeatedly drawn from a known population.

In ◘ Fig. 2.1, means of all the samples are not same; it's quite obvious also because these means are computed from distinct sets of data drawn from the population. The next question is whether we can find the mean and variance of these sample means.

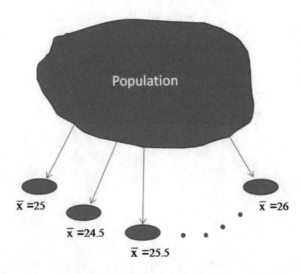

◘ **Fig. 2.1**    Drawing all possible samples of the same size from the population

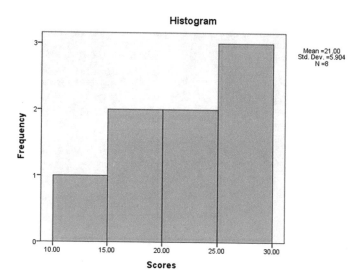

**Histogram**

Mean =21.00
Std. Dev. =5.904
N =8

□ **Fig. 2.2** Histogram of the population scores

In other words, what would be the distribution of sample mean? Let us first examine the distribution of means graphically. Consider the population denoted by $X$ as shown below.

$$X : 12 \quad 15 \quad 17 \quad 21 \quad 22 \quad 26 \quad 27 \quad 28.$$

The mean and standard deviation of this finite population can be obtained as follows:

$$\mu = 21$$
$$\sigma = 5.52$$

The distribution of $X$ in the population can be obtained by plotting the histogram as shown in □ Fig. 2.2. It can be seen that this distribution is negatively skewed. Let us examine the distribution of samples now. If all the possible samples of size 2 are drawn with replacement from this population, then such samples can be listed as shown in □ Table 2.1. Let us compute the mean of each of these samples as shown in □ Table 2.2. By computing the mean and standard deviation of these sample means, we get

$$\bar{x} = 21$$
$$s = 3.91$$

If the histogram is constructed for these sample means, it will look like as shown in □ Fig. 2.3.

It can be seen from □ Fig. 2.3 that the distribution of sample mean seems to be approaching towards normal distribution. Although the population distribution is highly skewed, the distribution of sample means even of size 2 is close to the normal distribution. As per the central limit theorem, whatever may be the distribution of the population but if the sample size is 30 or more then the distribution of the sample mean shall be approximately normally distributed.

**◘ Table 2.1**    List of all possible samples of size 2 drawn with replacement

| | | | | | | | |
|---|---|---|---|---|---|---|---|
| 12,12 | 12,15 | 12,17 | 12,21 | 12,22 | 12,26 | 12,27 | 12,28 |
| 15,12 | 15,15 | 15,17 | 15,21 | 15,22 | 15,26 | 15,27 | 15,28 |
| 17,12 | 17,15 | 17,17 | 17,21 | 17,22 | 17,26 | 17,27 | 17,28 |
| 21,12 | 21,15 | 21,17 | 21,21 | 21,22 | 21,26 | 21,27 | 21,28 |
| 22,12 | 22,15 | 22,17 | 22,21 | 22,22 | 22,26 | 22,27 | 22,28 |
| 26,12 | 26,15 | 26,17 | 26,21 | 26,22 | 26,26 | 26,27 | 26,28 |
| 27,12 | 27,15 | 27,17 | 27,21 | 27,22 | 27,26 | 27,27 | 27,28 |
| 28,12 | 28,15 | 28,17 | 28,21 | 28,22 | 28,26 | 28,27 | 28,28 |

**◘ Table 2.2**    Means of all the samples

| | | | | | | | |
|---|---|---|---|---|---|---|---|
| 12 | 13.5 | 14.5 | 16.5 | 17 | 19 | 19.5 | 20 |
| 13.5 | 15 | 16 | 18 | 18.5 | 20.5 | 21 | 21.5 |
| 14.5 | 16 | 17 | 19 | 19.5 | 21.5 | 22 | 22.5 |
| 16.5 | 18 | 19 | 21 | 21.5 | 23.5 | 24 | 24.5 |
| 17 | 18.5 | 19.5 | 21.5 | 22 | 24 | 24.5 | 25 |
| 19 | 20.5 | 21.5 | 23.5 | 24 | 26 | 26.5 | 27 |
| 19.5 | 21 | 22 | 24 | 24.5 | 26.5 | 27 | 27.5 |
| 20 | 21.5 | 22.5 | 24.5 | 25 | 27 | 27.5 | 28 |

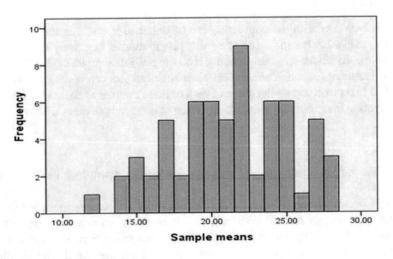

**◘ Fig. 2.3**    Histogram of the sample means

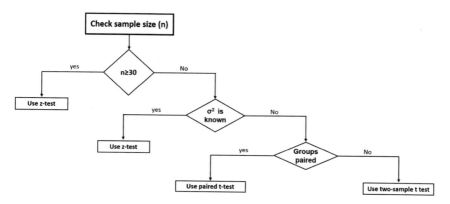

**Fig. 2.4** Decision criteria for using statistical test

We know that if samples are drawn from a normal population $N (\mu, \sigma^2)$, then the distribution of sample means ($\bar{x}$) will also follow normal distribution with mean $\mu$ and variance $\sigma^2/n$. This statement can be illustrated with the above example. Here, the mean of all sample means ($\bar{x}$), i.e. 21 is equal to the population mean $\mu$ (=21). Similarly, standard deviation of all the sample means is also equal to $\frac{\sigma}{\sqrt{n}}$, i.e.

$$s = \frac{\sigma}{\sqrt{n}} = \frac{5.52}{\sqrt{2}} = 3.91$$

Often there is a confusion among the researchers that the Z-test is meant for large samples ($n \geq 30$) and $t$-test is meant for small samples ($n < 30$). The central limit theorem justifies the use of Z-test in a large sample but the use of $t$-test depends upon whether the population variance is known or not. Irrespective of whether the population variance is known or unknown, in small sample it is mandatory that the sample must have been drawn from the normal population. In case of violation of normality assumption, non-parametric test such as Mann–Whitney $U$ should be used instead of $t$-test. The selection criteria for using Z- or $t$-test have been shown graphically in  Fig. 2.4.

### Estimating Parameters

We know that if the marks ($X$) obtained by the students in a test are normally distributed with mean $\mu$ and variance $\sigma^2$, then as per the area property of normal distribution we can make the following statement:

"We are confident that 95% scores would lie in between $[\mu - Z_{0.05/2}\sigma]$ and $[\mu + Z_{0.05/2}\sigma]$" ( Fig. 2.5).

The interpretation of this statement is that if 100 students appear in the exam then 95 students' marks would be in between $\bar{x} - 1.96\sigma$ and $\bar{x} + 1.96\sigma$, provided marks are normally distributed.

Let us consider that the blood sugar ($x$) is normally distributed with mean ($\mu$) and standard deviation ($\sigma$) being 90 mmol/l and 10 mmol/l, respectively. Then as per the normal distribution the areas would be distributed as shown in  Fig. 2.6.

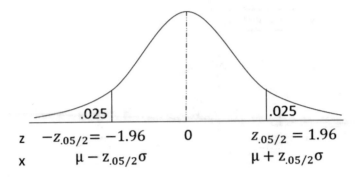

**⬛ Fig. 2.5**  Normal distribution showing area in tails

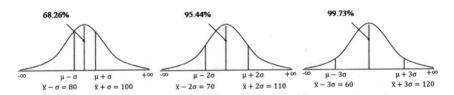

**⬛ Fig. 2.6**  Distribution of area in the normal distribution

With the above distribution of scores, the following three statements can be made:

a.  68.26% subject's sugar level is in between 80 and 100.
b.  95.44% subject's sugar level is in between 70 and 110.
c.  99.73% subject's sugar level is in between 60 and 120.

### Estimating Population Mean

In survey studies, either population mean or proportion is being estimated from the sample. As per the central limit theorem, we know that if a sample is drawn from the population having mean $\mu$ and variance $\sigma^2$ then the sample mean, $\bar{x}$, follows normal distribution with mean $\mu$ and variance $\sigma^2/n$ provided sample size is 30 or more and we can apply the normal distribution property in estimating the population mean. Thus, in large sample ($n \geq 30$) we are 95% confident that the confidence interval

$$\bar{x} - Z_{0.05/2} \frac{\sigma}{\sqrt{n}} \quad \text{to} \quad \bar{x} + Z_{0.05/2} \frac{\sigma}{\sqrt{n}}$$

will include population mean. In other words, if 100 such samples are drawn, then at least 95 times this confidence interval will include population mean. What happens if a random sample of size less than 30 is drawn from the normal population $N(\mu, \sigma^2)$ where $\sigma^2$ is unknown? In that case, the $\bar{x}$ follows $t$-distribution having mean $\mu$ and variance $S^2/n$. In such situation, 95% confidence interval shall be obtained by

$$\bar{x} - t_{0.05/2, n-1} \frac{S}{\sqrt{n}} \quad \text{to} \quad \bar{x} + t_{0.05/2, n-1} \frac{S}{\sqrt{n}} \tag{2.1}$$

On the basis of the discussion mentioned above, let us see how confidence level and confidence interval are defined.

### Confidence Level

Confidence level is an indicator of accuracy in repeating a statistical test. Let us assume that the IQ of the students is normally distributed with mean $\mu$ and standard deviation 6. If a random sample of size 36 is drawn from this population and the sample mean is 75, then 95% confidence interval for population mean would be

$$\bar{x} - 1.96\frac{\sigma}{\sqrt{n}} \text{ to } \bar{x} + 1.96\frac{\sigma}{\sqrt{n}}$$

or

$$75 - 1.96\frac{6}{\sqrt{36}} \text{ to } 75 + 1.96\frac{6}{\sqrt{36}}$$

or

$(73-77)$ approximately.

It simply indicates that if random samples are drawn 100 times, at least 95 times the confidence intervals would include the population mean. This 95% is the confidence level. This fact can be shown in ◘ Fig. 2.7 in which 9 out of 10 confidence intervals include population mean. Thus, the confidence level in this case is 90%. The confidence level can also be termed as confidence coefficient; the only difference is that the confidence coefficient is represented in fraction, whereas confidence level is represented in percentage.

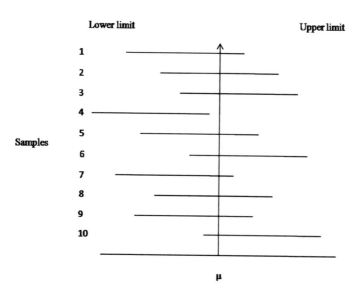

◘ **Fig. 2.7** Confidence intervals constructed on the basis of the sample for estimating population mean

**2**

> **Definition**
> **Confidence level** can be defined as the degree of certainty that the true value of population parameter lies within the confidence interval.

## Confidence Interval

Confidence interval can be defined as the limits within which the population parameters such as mean or proportion are supposed to lie with some confidence. We have seen that if a variable $x$ is normally distributed with mean $\mu$ and variance $\sigma^2$ then the confidence interval of mean with $(1 - \alpha) \times 100\%$ confidence can be obtained as

$$\bar{x} - Z_{\alpha/2}\frac{\sigma}{\sqrt{n}} \text{ to } \bar{x} + Z_{\alpha/2}\frac{\sigma}{\sqrt{n}} \tag{2.2}$$

> **Definition**
> A **confidence interval** is the estimated range of values obtained from the sample data which is likely to include the unknown population parameter.

We shall investigate the factors that affect the confidence interval by means of an example. In estimating subjects' height in the population, let us assume that a sample of 36 subjects is randomly drawn from the population and their heights are measured. Mean height of the sample is 170 cm, and the population standard deviation is 10 cm. By using the Formula (2.2), we can construct confidence intervals of mean having different confidence levels.

*90% Confidence Interval for Estimating Mean*

$$\bar{x} - Z_{0.1/2}\frac{\sigma}{\sqrt{n}} \text{ to } \bar{x} + Z_{0.1/2}\frac{\sigma}{\sqrt{n}}$$

$$170 - 1.645\frac{10}{\sqrt{36}} \text{ to } 170 + 1.645\frac{10}{\sqrt{36}}$$

$$(167.26 - 172.74)$$

*95% Confidence Interval for Estimating Mean*

$$\bar{x} - Z_{0.05/2}\frac{\sigma}{\sqrt{n}} \text{ to } \bar{x} + Z_{0.05/2}\frac{\sigma}{\sqrt{n}}$$

$$170 - 1.96\frac{10}{\sqrt{36}} \text{ to } 170 + 1.96\frac{10}{\sqrt{36}}$$

$$(166.73 - 173.27)$$

CI  167.26 172.74    166.73  173.27    165.7  174.3

◻ **Fig. 2.8** Confidence intervals with different confidence levels

*99% Confidence Interval for Estimating Mean*

$$\bar{x} - Z_{0.01/2}\frac{\sigma}{\sqrt{n}} \text{ to } \bar{x} + Z_{0.01/2}\frac{\sigma}{\sqrt{n}}$$

$$170 - 2.58\frac{10}{\sqrt{36}} \text{ to } 170 + 2.58\frac{10}{\sqrt{36}}$$

$$(165.7 - 174.3)$$

The above three computed confidence intervals can be shown graphically in ◻ Fig. 2.8.

### Factors Affecting Confidence Interval

a.  *Confidence level*: By looking at ◻ Fig. 2.8, you can see that if the confidence level increases the confidence interval also increases. Thus, the confidence interval depends upon the degree of confidence level you wish to have in estimating the population mean.

b.  *Sample size:* In Eq. (2.2) if the sample size n increases the confidence interval decreases. Thus, the larger the sample, the narrower the confidence interval in estimating the population mean. In other words, the larger sample provides more accuracy in estimation.

c.  *Population standard deviation:* From Eq. (2.2), it can be seen that the larger population standard deviation tends to increase the confidence interval. In other words, more heterogeneity in the population wider is the confidence interval and vice versa.

### Estimating Population Proportion

In most of the survey studies, we are interested in estimating population proportion of any characteristics. For instance, we may wish to estimate the proportion of non-vegetarians in a city or we may like to know what proportion of population is smokers. In such situations, we can estimate the proportion of the characteristics by knowing its distribution.

If $x$ is a variate, occurrence of which is a success and non-occurrence is a failure, then the sample proportion of success $\hat{p}$ follows binomial distribution. But in a large sample the distribution of a sample proportion $\hat{p}$ follows normal distribution with mean $p$ and variance $pq/n$, where $p$ is the population proportion and $q = 1 - p$. Even if the sample size is less than 30 the distribution of sample proportion $\hat{p}$ follows normal distribution provided $n.\hat{p} \geq 5$ and $n.\hat{q} \geq 5$.

Thus, in large sample ($n \geq 30$) we are 95% confident that the confidence interval

$$\hat{p} - Z_{0.05/2}\sqrt{\frac{p \times q}{n}} \text{ to } \hat{p} + Z_{0.05/2}\sqrt{\frac{p \times q}{n}} \qquad (2.3)$$

will include the population proportion. The sample size for proportion should always be greater than 30, i.e. large, never small.

## Hypothesis Testing

In hypothesis testing experiment(s), research hypothesis is tested by negating the null hypothesis. For instance, if it is desired to test whether average IQ of college students is greater than 95, then we shall construct a research hypothesis as $H_1: \mu > 95$. In order to test this hypothesis, we shall construct a null hypothesis, $H_0: \mu = 95$. On the basis of the sample, if $H_0$ is rejected we accept the research hypothesis. The research hypothesis is also termed as alternative hypothesis. Many times when a researcher is asked about his hypothesis in the study he refers to the null hypothesis; instead, it is the research hypothesis which he intends to test. The null hypothesis is simply a via media for testing the research hypothesis.

Usually, in hypothesis testing experiments, we are interested in testing the population mean or proportion to any predefined value. However, in different situations we are interested in testing different parameters such as difference of two group means, equality of variances, significance of correlation coefficient and regression coefficient, etc.

Let us understand the concepts involved in hypothesis testing experiment. We know that if a random sample is drawn from a normal population with mean $\mu$ and variance $\sigma^2$ then the distribution of sample means $\bar{x}$ also follows normal distribution with mean $\mu$ and variance $\sigma^2/n$. Thus, by owing to the area property of normal distribution if a random sample is drawn from the population there are 95% chances that its mean will fall within the following interval:

$$\mu - 1.96\frac{\sigma}{\sqrt{n}} \text{ to } \mu + 1.96\frac{\sigma}{\sqrt{n}}$$

If the sample mean is less than $\mu - 1.96\sigma/\sqrt{n}$ or greater than $\mu + 1.96\sigma/\sqrt{n}$, we may infer that the sample does not belong to the population with 5% error and if it falls within these two limits we may infer that the sample belongs to the population.

Consider again the IQ testing experiment in which it is desired to know whether the IQ of college students differs from 95 or not. Let us suppose that a sample of 36 students is randomly selected which has a mean IQ as 99. Further, it is known that the population standard deviation of IQ is 12. Can it be concluded that the sample does not belong to the population having mean 95?

Here we are interested to test the null hypothesis $H_0: \mu = 95$ against the alternative hypothesis $H_1: \mu \neq 95$. As per the central limit theorem, sample mean $\bar{x}$ follows normal distribution with mean 95 (population mean) and standard deviation 2 ($=12/\sqrt{36}$). Let

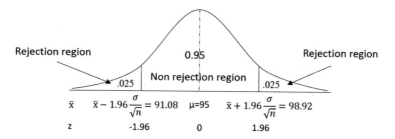

**□ Fig. 2.9** Critical region in a hypothesis testing experiment

us see whether we can conclude that the sample belongs to the population. For a sample to belong to the population with 95% confidence, its mean should lie in between 91.08 (95 − 1.96 × 2) and 98.92 (95 + 1.96 × 2) as shown in □ Fig. 2.9. Since sample mean is 99 which is beyond the upper limit 98.02, hence, it may be concluded that the sample does not belong to the population and the null hypothesis is rejected.

Usually, hypothesis is tested by constructing a test statistic. This test statistic follows a particular sampling distribution under the assumption of null hypothesis, and on the basis of its distribution decision about rejecting or not rejecting the null hypothesis is taken. Test statistic is computed by the following generalized formula:

$$\text{Test statistic} = \frac{\text{Statistic} - \text{E(Statistic)}}{\text{SE(Statistic)}} \tag{2.4}$$

Here, statistic is the sample characteristic, $E$(statistic) is the expected value of the statistic which is equal to the population parameter and SE(statistic) is the standard error of the statistic. For large sample ($n \geq 30$), the test statistic follows normal distribution due to central limit theorem.

In the above-mentioned IQ testing illustration, $\bar{x}$, the sample mean is the statistic. Thus, the test statistic would be

$$Z = \frac{\bar{x} - \mu}{\sigma/\sqrt{n}} \tag{2.5}$$

This test statistic $Z$ follows standard normal distribution with mean 0 and standard deviation 1. Let us compute this $Z$ statistic for the above-mentioned illustration

$$Z = \frac{99 - 95}{12/\sqrt{36}} = 2$$

If the values of $Z$ fall within $-1.96$ to $1.96$, do not reject the null hypothesis and if it falls beyond these limits reject the null hypothesis and accept the alternative hypothesis with 95% confidence. Since calculated $Z$ is 2, which is greater than 1.96, hence, the null hypothesis is rejected and the sample cannot be considered to be drawn from the population having mean 95.

**2**

□ **Table 2.3**    Decision options in testing the null hypothesis

| Researcher's decision | | Actual state of $H_0$ | |
|---|---|---|---|
| | | $H_0$ *true* | $H_0$ *false* |
| *Reject $H_0$* | | Type I error ($\alpha$) (False positive) | Correct decision $(1 - \beta)$ |
| *Do not reject $H_0$* | | Correct decision | Type II error ($\beta$) (False negative) |

## Type I and Type II Errors

As we know that in hypothesis testing experiment research hypothesis is tested by negating the null hypothesis, hence, the focus of the researcher is to see whether the null hypothesis can be rejected in favour of research hypothesis or not on the basis of the sample. One should understand that the null hypothesis is never accepted, rather it is meant for rejection. In case the sample does not give any evidence for rejecting the null hypothesis, one should opine that the null hypothesis is failed to be rejected on the basis of the sample instead of accepting the null hypothesis.

Since the decision is taken on the null hypothesis, there are four courses of actions available to the researcher out of which two are correct and two are wrong. The two correct decisions are rejecting the null hypothesis when it is false and not rejecting the null hypothesis when it is true. On the other hand, two wrong decisions are rejecting null hypothesis when it is true and not rejecting the null hypothesis when it is false. All these four cases are summarized in □ Table 2.3.

These two wrong decisions of the researcher are known as statistical errors. Rejecting a null hypothesis when it is true is known as Type I error ($\alpha$) and not rejecting the null hypothesis when it is false is Type II error ($\beta$). Type I error is also termed as false positive as due to this error the researcher will accept the false claim. Similarly, Type II error is also known as false negative because the claim is correct but the researcher is unable to accept it due to this error. In hypothesis testing experiment, a researcher always tries to minimize both these errors simultaneously. This is possible by increasing the sample size because minimizing one type of error enhances the other. Another approach is to see as to which error is more severe.

> **Definition**
> A Type I error can be defined as the probability of rejecting the null hypothesis when it is true. It is represented by $\alpha$.

> **Definition**
> A Type II error is the probability of not rejecting the null hypothesis when it is false. It is represented by $\beta$.

| Table 2.4 | Decision options in drug testing experiment |

| Researcher's decision | | Actual state of $H_o$ | |
|---|---|---|---|
| | | $H_0$ true | $H_0$ false |
| | Reject $H_0$ | Type I error ($\alpha$) (wrongly accepting ineffective drug) | Correct decision |
| | Do not reject $H_0$ | Correct decision | Type II error ($\beta$) (wrongly rejecting the effective drug) |

Consider an experiment in which a drug is tested for its efficacy. Here, the null hypothesis $H_0$ is that the drug is not effective which is tested against the alternative hypothesis $H_1$ that the drug is effective. If Type I error is committed, the researcher will wrongly reject the null hypothesis $H_0$ in favour of alternative hypothesis. In other words wrongly rejecting, the null hypothesis will force the researcher to accept the wrong claim which is serious in nature. On the other hand, in Type II error the researcher does not reject the null hypothesis when the claim about the drug is correct. Implications of these two errors have been shown in ☐ Table 2.4.

The Type I error can be considered as consumer's risk, whereas Type II error is a producer's risk. Since Type I error is more serious than the Type II error in this case, hence, it is always fixed at an acceptable low level.

Level of significance is the probability with which we reject a null hypothesis which is true, and Type I error is the amount of error that we can at best tolerate. The level of significance is denoted by $\alpha$. Similarly, probability of committing Type II error is represented by $\beta$. Thus, we can write

$\alpha$ = Probability of rejecting null hypothesis when it is true
$\beta$ = Probability of not rejecting null hypothesis when it is false.

Usually, Type I error ($\alpha$) is taken as 0.05 or 0.01, whereas Type II error ($\beta$) is kept at 0.2 or less in the study. Depending upon the implications of the study, Type I error can be further reduced.

**Power of the Test**
Power of the test is obtained by $1 - \beta$. It is the probability of rejecting the null hypothesis when null hypothesis is false. It means that if the drug is effective the test should accept it.

Thus, Power = $1 - \beta$ = Prob. (Rejecting the null hypothesis $H_0$ when $H_0$ is false)

**Definition**
Power of the test can be defined as the probability of correctly rejecting the null hypothesis. It is computed by $1 - \beta$.

2

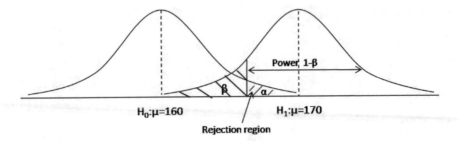

Power, 1-β; β; α; H₀:μ=160; H₁:μ=170; Rejection region

**□ Fig. 2.10**    Distribution of mean under null and alternative hypotheses

The next question is how much power one should have in the experiment. Logically, one should have at least 0.8 power. This means that 80% of the time, the test should reject the null hypothesis if the drug is effective. If the power in any hypothesis testing experiment is 50% or less, then there is no meaning of performing the test. Simply tossing a coin and deciding the effectiveness of the drug would be sufficient.

### Relationship between Type I and Type II Errors

We have seen that Type I and Type II errors are linked with one another. If one decreases other will increase but not in direct proportion. We shall investigate their relationship by means of an illustration. Let $x$ denotes the height which is normally distributed with unknown mean $\mu$ and standard deviation 14. If a researcher draws a random sample of 36 students for testing the null hypothesis $H_0$: $\mu = 160$ against $H_1$: $\mu > 160$ at 5% level, then let us see what would be the power in the test. We know that under the null hypothesis sample mean follows normal distribution. In order to find the distribution of $\bar{x}$ under $H_1$, the value of population mean needs to be specified. Suppose we wish to test the above-mentioned null hypothesis $H_0$: $\mu = 160$ against the alternative hypothesis $H_1$: $\mu = 170$. Then under $H_0$ and $H_1$ the distribution of $\bar{x}$ shall be as shown in □ Fig. 2.10.

Here, $\alpha$ is the probability of rejecting $H_0$ when it is true. The null hypothesis $H_0$ is rejected if the test statistic falls in the rejection region as shown by the shaded area marked with $\alpha$ □ Fig. 2.10. On the other hand, $\beta$ is the probability of not rejecting $H_0$ when $H_1$ is true and this is indicated by the shaded area marked with $\beta$ in the figure. The power is the remaining area of the normal distribution when $H_1$ is true. It can be seen from □ Fig. 2.10 that if $\alpha$ decreases, $\beta$ increases with the result power reduces. It can also be seen from the figure that the amount of decrease in $\alpha$ is not the same as the amount of increase in $\beta$. It is interesting to note that as the Type I error ($\alpha$) increases the power of the test ($1 - \beta$) also increases and vice versa.

Often researchers are amazed if their null hypothesis is rejected at the significance level of 0.01 instead of 0.05 and they feel that their results are more powerful. If that is the case, why not to reduce $\alpha$ to zero? In that case the null hypothesis will never be rejected; howsoever, the claim is correct. Similarly, if the $\beta$ is taken as zero then every time null hypothesis would be rejected in favour of the claim. So, what should be the strategy? Since we have seen above that as $\alpha$ reduces, power of the test also reduces,

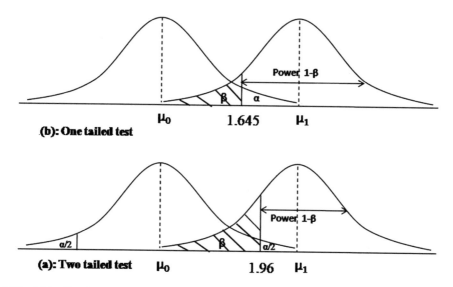

■ **Fig. 2.11** Showing comparison of power in one- and two-tailed tests

hence, a perfect balance is required in fixing $\alpha$ and $\beta$ in the experiment. We shall discuss more about these issues in the later chapters.

### One-Tail and Two-Tail Tests

While estimating sample size in a hypothesis testing experiment, it is important to know whether one-tail or two-tail tests are used. In one-tailed test entire critical region, $\alpha$ is shifted in one tail, whereas in two-tailed test $\alpha$ is divided in both the tails. Due to this, in one-tailed test, $\beta$ becomes less in comparison to that of two-tailed test for same $\alpha$, resulting in more power $(1 - \beta)$ in one-tailed test in comparison to the two-tailed test. This fact can be seen in ■ Fig. 2.11a, b. Thus, the power increases at the cost of $\alpha$. In other words, one-tailed test is more powerful than the two-tailed test for the same level of $\alpha$.

---

#### Definition
A one-tailed test can be defined as a test for statistical hypothesis in which the rejection region is only on one side of the sampling distribution.

---

#### Definition
A two-tailed test can be defined as a test for statistical hypothesis in which the rejection region is on both sides of the sampling distribution.

## Procedure in Hypothesis Testing Experiment

In hypothesis testing experiments, several boundary conditions need to be fixed along with the desired power without which testing results have no meaning. Following are the steps involved in it:

a.  Write null and alternative hypotheses. Alternative hypothesis is the one in which the researcher is interested to test. It may be one-tailed or two-tailed depending upon the knowledge about the effect. If the effect is known to be one directional, one-tailed test should be used; otherwise, two-tailed test may be preferred. For instance, one-tailed test may be used if we are interested to test the effect of exercise intervention on $VO_2$ max (a way to measure the efficiency of heart) in a 3-week programme. Because it is a known fact that the exercise intervention will only improve the $VO_2$ max but will not deteriorate in general. Here, the issue is whether increase is significant or not. But if the effect of a new drug is to be seen on some physiological parameter and the effect is not known from earlier studies or there are contradictory views about its effect, in that situation two-tailed test may be preferred.

b.  Decide whether to use one-tailed or two-tailed test.

c.  Fix the level of significance ($\alpha$). Usually, $\alpha$ should be any value in between 0.01 and 0.10 depending upon the severity of Type I error. Normally, it is taken as 0.05 or 0.01.

d.  Fix the Type II error ($\beta$). Since power is $1 - \beta$ and it should be at least 0.8 or more, hence, $\beta$ should be fixed at 0.2 or less. Any test having power 50% or less is useless because in that case effect may be tested by simply tossing a coin instead of hypothesis testing experiment.

e.  Decide minimum detectable difference ($d$). The researcher should decide as to how much minimum effect is required to be detected in the hypothesis testing experiment. For instance, in a 4-week weight management programme for how much reduction in mean weight of the participants can be considered for the programme to be effective. One can wish to detect the reduction of 2, 3 or 4 kg on an average in the programme. If the researcher does not decide this at the outset and the weight management programme results in reduction of only 200 g on an average, can the programme be considered to be effective?

f.  Find the population variance ($\sigma^2$) of the parameter whose effect is to be investigated. For instance, if the effect of exercise intervention on $VO_2$ max is to be seen among the men in the age category 40 to 50 years, then one should know the variance of $VO_2$ max of the men aged 40–50 years in the population. In fact, population variance is rarely known; hence, its estimate $S^2$ is either obtained from the similar studies conducted earlier or may be determined from the pilot study. The readers should note that if the estimate $S^2$ is obtained from the pilot study, then its data should not be used for the main study.

g.  Determine the estimated sample size, $n$, for the given value of $\alpha$, $\beta$, $S^2$, minimum detectable difference ($d$) and the type of test (one tail or two tail). The manual procedure of estimating the required sample size has been discussed for few situations in ▶ Chaps. 3 and 4. However, for the applied researchers, these methods have been discussed in ▶ Chaps. 6 and 7 by using the G*Power software.

h. With the estimated sample size $n$ obtained in step ($g$) compute test statistic, say, $t$.

i. If null hypothesis is rejected at $\alpha$ significance level, then one can say that the treatment is effective with at least $d$ difference in their mean value and the test has power $(1 - \beta)$.

## Effect Size

Effect size is equivalent to minimum detectable difference in a hypothesis testing experiment. It is a method of quantifying the difference between the two groups' means. Since different variables have different units, hence, in order to standardize, the effect size is obtained by the following formula:

$$\frac{\text{Mean of experimental group} - \text{Mean of control group}}{\text{Standard deviation}}$$

**Definition**
Effect size can be defined as the strength of relationship between the intervention and the variable on which its effect is investigated. It determines the practical significance of the effect in the experiments.

As per Cohen (1988), the power should be at least 0.8, i.e. in four out of five cases the null hypothesis should be correctly rejected. Power more than 0.8 in the study is always welcome but this will increase sample size as well and may lead to the illusion of seeing an effect falling finally into a false-positive trap. Indeed, the more the trials one performs, the more likely to incur a type one error (Rüttimann and Wegener 2015). In determining the sample size for a specific power $(1 - \beta)$, the researcher needs to specify the minimum detectable difference which he wishes to test. Cohen's definition of small (0.2), medium (0.5) and large (0.8), effects can be helpful in such effect size specifications (Smith and Bayen 2005). However, researchers should understand the fact that these conventions may have different meanings for different tests (Erdfelder et al. 2005). In order to compute effect size, the population variability should be known or obtained by estimation. If the researcher is not able to decide how much detectable difference he desires to investigate and population variance is an issue, then Cohen's guidelines may be adopted for the effect size. However, in biological studies where maximum control is observed in the experiment, the assessment of effect size is based on higher values. A comparative effect of sizes and their significance are listed in ◘ Table 2.5.

In any hypothesis testing experiment '$p$' value signifies that the effect exists, whereas the effect size reveals the size of the effect. For instance, if the hypothesis test reveals that the drug A is more effective than B in reducing the blood pressure at the 0.05 significance level, we may understand that the drug A reduces blood pressure more than B but the effect size explains how much it reduces in comparison to B.

2

| Magnitude | Effect size | |
|---|---|---|
| | Normal experiment | Biological experiment |
| Low | 0.2 | 0.5 |
| Medium | 0.5 | 1.0 |
| High | 0.8 | 1.5 |

◻ Table 2.5   Effect size in studies

## Summary

One of the assumptions for all parametric tests is normality. As per the central limit theorem, sample mean is normally distributed even if the population is non-normal, provided the sample is large. The confidence level indicates the probability that an interval will include population characteristics. On the other hand, confidence interval is a limit which includes population parameter with some confidence. The research hypothesis may be tested by using the concept of confidence interval or by testing the null hypothesis. The null hypothesis is tested by using a test statistic which is constructed by dividing the difference between statistics and its expected value by the standard error of the statistic. In testing null hypothesis, the researcher needs to control two types of statistical errors, Type I and Type II. Type I error ($\alpha$) indicates rejecting null hypothesis when it is true, whereas Type II error ($\beta$) refers to retaining the null hypothesis when alternative hypothesis is true. Power of the test is the probability of rejecting null hypothesis when alternative hypothesis is true. It is computed by $1 - \beta$. The two statistical errors are reciprocal to each other. Thus, if Type I error decreases, power also decreases and vice versa. Besides controlling both the errors in hypothesis testing experiments, one needs to decide as to how much effect is required to be tested as well. The effect size can be defined as the ratio of the minimum detectable difference and standard deviation of the measure.

- Exercises
  1. What is hypothesis testing?
  2. Comment on the statement, "$z$-test is used for large sample whereas $t$-test is used for small sample". Do you require data to come from the normal distribution for using $t$-test?
  3. What do you mean by the confidence level and confidence interval? Discuss with some examples.
  4. How can you reduce the confidence interval for estimating population mean in a research study?
  5. What is test statistics and how it is formed? Why for all the parametric tests normality of data is required?
  6. What do you mean by statistical errors? Which of the two errors; Type I or Type II is more serious and why? Explain by means of an example. What happens to power if Type I error is increased or decreased?

7. What is the significance of power in hypothesis testing? What happens if the power of the test is not fixed in advance? How power of a test is affected by the effect size and sample size?

8. If $t$-test is significant at 1% level instead of 5% level, researcher usually considers the findings as more accurate. Do you agree with this concept? In either case defend your viewpoint with arguments.

9. Describe the steps used in hypothesis testing experiment.

10. Explain the meaning of effect size and its practical significance in hypothesis testing experiment. What is the difference between the effect size and minimum detectable difference?

11. If samples of size 2 are drawn from the population values denoted by X: 1, 2, 3, 4, 5, then show that the mean and variances of the sample means shall be $\mu$ and $\sigma^2/n$, respectively, where $\mu$ and $\sigma^2$ are the mean and variance of the population, respectively.

12. Show that the mean is an unbiased estimate of the population. Prove it either by formula or by a set of empirical data.

13. For the population of elements S: [16, 18, 20]

   a. List all possible samples of size two, chosen by simple random sampling with replacement.

   b. Calculate the variance of the population.

   c. Calculate the variance of the sample means.

14. If mean of the systolic pressure data of a randomly drawn sample of size 36 is 130 and standard deviation of the systolic pressure in the population is 12, construct 95% and 99% confidence intervals of population mean.

15. While producing bulbs by a manufacturing company, a random sample of size 100 consists of 15 defective bulbs. What will be the 95% confidence interval for estimating the population proportion of defectives?

16. A manufacturer produces basketball for retail stores. It is known that the basketball diameter is normally distributed and has a standard deviation of 0.04 inches. A random sample of 25 basketball has mean diameter, $\bar{x} = 9.51$ inches. Construct a 95% confidence interval population mean of basketball diameter.

17. The average weight of a random sample of 100 people from a city is 55 kg. It is known that the weight of the population is a random variable which follows a normal distribution having variance 9 kg. Construct 99% confidence interval for population mean of weights.

■ **Answer**

13. a. 16,16
      16,18
      16,20
      18,16
      18,18
      18,20
      20,16
      20,18
      20,20

 b. Variance of the population $= 2.67$
 c. Variance of the sample means $= 1.33$
14. 95% CI: {126.08 to 133.92}
    99% CI: {124.84 to 135.16}
15. 95% CI: {0.08 to 0.22}
16. 95% CI: {9.49 to 9.53}
17. 99% CI: {52.94 to 57.06}

# Bibliography

Altman, D. G. (1990). Practical statistics for medical research. CRC Press, Section 15.3B.

Cox, D. R. (2006). *Principles of statistical inference*. Cambridge: Cambridge University Press. ISBN 978-0-521-68567-2.

Fisher, Sir R. A. (1956) (1935). Mathematics of a lady tasting tea. In J. R. Newman (Ed.), *The world of mathematics* (Vol. 3) [Design of Experiments]. Courier Dover Publications. ISBN 978-0-486-41151-4. Originally from Fisher's book Design of Experiments.

Gigerenzer, G. (2004). The null ritual what you always wanted to know about significant testing but were afraid to ask (PDF). The SAGE handbook of quantitative methodology for the social sciences, pp 391–408. ▶ https://doi.org/10.4135/9781412986311.

Halpin, P. F, & Stam, H. J. (Winter 2006). Inductive inference or inductive behavior: Fisher and Neyman: Pearson approaches to statistical testing in psychological research (1940–1960). *The American Journal of Psychology, 119*(4), 625–653. JSTOR 20445367. PMID 17286092. ▶ https://doi.org/10.2307/20445367.

Lehmann, E. L., & Romano, J. P. (2005). Testing statistical hypotheses, 3rd edn. New York: Springer. ISBN 0-387-98864-5.

Neyman, J., & Pearson, E. S. (1967) (1933). The testing of statistical hypotheses in relation to probabilities a priori. Joint Statistical Papers. Cambridge University Press, pp 186–202.

Shields, P. M., & Rangarajan, N. (2013). *A playbook for research methods: Integrating conceptual frameworks and project management* (pp. 109–157). Stillwater, OK: New Forums Press.

Fisher, R. A. (1925). *Statistical methods for research workers* (p. 43). Edinburgh: Oliver and Boyd.

Sokal, R. D., & Rohlf, F. J. (1969). *Biometry: The Principles and practice of statistics in biometric research*. San Francisco: Freeman.

Newcombe, R. G. (1998). Two sided confidence intervals for the single proportion: Comparison of seven methods. *Statistics in Medicine, 17,* 857–872.

Larsen, R. J., & Stroup, D. F. (1976). *Statistics in the real world: A book of examples*. Macmillan. ISBN 978-0023677205.

Rüttimann, B., & Wegener, K. (2015). *Introduction to lean manufacturing and six sigma quality control*. ETH Tools-IV Kurs, Lecturing notes HS2015, D-MAVT.

Jung, S. H., & Ahn, C. (2000). Estimation of response probability in correlated binary data: A new approach. *Drug Information Journal, 34,* 599–604.

Sheskin, D. (2004). *Handbook of parametric and nonparametric statistical procedures*. CRC Press. ISBN 1584884401.

Smith, R. E., & Bayen, U. J. (2005). The effects of working memory resource availability on prospective memory: A formal modeling approach. *Experimental Psychology, 52*(4), 243–256. ▶ https://doi.org/10.1027/1618-3169.52.4.243.

Sotos, A. E. C., Vanhoof, S., Van den Noortgate, W., & Onghena, P. (2009). How confident are students in their misconceptions about hypothesis tests? *Journal of Statistics Education, 17*(2).

Steel, R. G. D., & Torrie, J. H. (1960). *Principles and procedures of statistics with special reference to the biological sciences* (p. 350). McGraw Hill.

# Understanding Concepts in Estimating Sample Size in Survey Studies

© Springer Nature Singapore Pte Ltd. 2020
J. P. Verma and P. Verma, *Determining Sample Size and Power in Research Studies*,
https://doi.org/10.1007/978-981-15-5204-5_3

- **Learning Objectives**

After going through this chapter, the readers should be able to

- Compute sample size in estimating population mean and proportion in survey research.
- Explain as to how effect size, power, significance level and variability affect determining sample size in survey research.
- Determine sample size in estimating difference between two population means.

## Introduction

A small sample yields inaccurate findings, conversely a large sample is an unnecessary mobilization of extra resources, therefore, determining optimum sample size is a crucial exercise in research studies. Moreover, in large sample even small effect may be found to be significant which may not have any practical utility. There are two concepts which will be conceptualized and its application will be discussed for inferring the optimal sample size. It is broadly defined as precision of estimates and power of the test. The concept of precision is used in determining sample size in survey studies, whereas in hypothesis testing experiments, sample size is estimated on the basis of power required in the study. In this chapter, we will discuss the theoretical foundations to arrive at the formula for calculating optimal sample size based on the concept of precision. The following chapter will deal with the concept of power, which is used in hypothesis testing.

## Determining Sample Size in Estimating Population Mean

We know that if samples of size $n$ are drawn from the normal population, $N(\mu, \sigma^2)$ then sample mean $\bar{x}$ is also normally distributed with mean $\mu$ and variance $\sigma^2/n$. Thus, $(1 - \alpha) \times 100\%$ confidence limits for the population mean are computed by

$$\bar{x} - Z_{\alpha/2}\frac{\sigma}{\sqrt{n}} \text{ to } \bar{x} + Z_{\alpha/2}\frac{\sigma}{\sqrt{n}}$$

In other words, if $\alpha = 0.05$ then we are 95% confident that these two limits will include population mean which is being estimated from the sample data. The next question is determination of the precision we want in estimating the populations mean. For instance, if we wish to have our estimate within $\pm d$ of the true population mean as shown in ◻ Fig. 3.1, then let us see how much sample size we would require to have this much precision. In estimating population mean, $d$ is the amount of tolerable variation, whereas in estimating population proportion $d$ represents the level of error which can be tolerated in estimating population proportion. For the error, '$d$' on either side of the mean

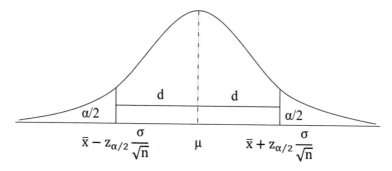

■ **Fig. 3.1** $(1 - \alpha) \times 100\%$ confidence limits for the population mean

$$2d = \left(\bar{x} + Z_{\alpha/2}\frac{\sigma}{\sqrt{n}}\right) - \left(\bar{x} - Z_{\alpha/2}\frac{\sigma}{\sqrt{n}}\right)$$

$$d = Z_{\alpha/2}\frac{\sigma}{\sqrt{n}}$$

$$n = Z_{\alpha/2}^2\left(\frac{\sigma}{d}\right)^2 \tag{3.1}$$

By using Eq. (3.1), the sample size in estimating mean of any population character-istics can be obtained. Here, $\alpha$ is the level of significance which depends upon how much confidence one wishes to have in estimation. If 95% confidence is required, then $Z_{\alpha/2}$ would be 1.96. The value of $Z$ for other values of $\alpha$ can be obtained from the standard normal distribution table.

The value of $\sigma$ is obtained from similar studies conducted earlier. It can be obtained from the review of literature. Here, $d$ is the amount of error which one can tolerate in estimating mean.

In case the population variance, $\sigma^2$ is not known and is being estimated by $S$ from similar studies conducted earlier, then instead of $Z_{\alpha/2}$ the value of $t_{\alpha/2,\nu}$ is used to esti-mate the sample size. In such situation, the sample size is determined by iteration pro-cess using the formula (3.2). The $\nu$ is the degrees of freedom obtained by $n - 1$.

$$n = t_{\alpha/2,\nu}^2\left(\frac{S}{d}\right)^2 \tag{3.2}$$

## Factors Affecting Sample Size

By looking at Eq. (3.1), it can be noticed that as the level of confidence increases the sample size also increases (for 95% confidence. $Z_{0.05/2} = 1.96$ and for 99% confidence. $Z_{0.01/2} = 2.58$). Thus, confidence level in estimating population mean increases with the increase in sample size, provided other factors remain constant. If variability of the population is more, the larger sample is required for the same level of confidence and precision. It is quite obvious; if you wish to have a precise estimate, even a small sample will be sufficient if the population is homogeneous. But if the population is heterogeneous, a larger sample is required. More precise estimate can be obtained by reducing $d$ which would require large sample for the same level of confidence and population variance.

## Sample Size Determination for Estimating Mean When Population SD Is Known

- Illustration 3.1

What would be the sample size in estimating mean height of male students with 95% confidence level from a sample of students that is within 4 cm of the true mean? Given that population standard deviation of height, $\sigma = 10$ cm.

- Solution

Given confidence level $= 95\%$, $Z_{0.05/2} = 1.96$, $\sigma = 10$ cm and $d = 4$ cm. Using Formula (3.1)

$$n = 1.96^2 \times (10/4)^2 \approx 24$$

Thus, 24 sample points are required to have 95% confidence estimate of population mean that is within 4 cm of the population mean.

## Sample Size Determination for Estimating Mean When Population SD Is Unknown

- Illustration 3.2

In illustration 3.1, if the population standard deviation of height is not known and is being estimated by the similar studies conducted earlier as $S = 8$ cm, what would be the estimated sample size?

- Solution

Since population standard deviation has been estimated, hence, we shall determine the sample size by iteration method using Formula (3.2). Here $S = 8$ cm, $d = 4$ cm.

Since the value of $t_{\alpha/2,v}$ cannot be determined until $n$ is known. Let us assume that the estimated sample size is 20 then $t_{0.05/2,19} = 2.093$.

$$n = \left(t_{\alpha/2,v}\right)^2 \times \left(\frac{S}{d}\right)^2$$
$$= 2.093^2 \times (8/4)^2 = 17.52$$

Our estimated value of $n = 20$ seems to be higher; hence, let us assume now $n = 18$, then $t_{0.05/2,17} = 2.110$.

$$n = 2.110^2 \times (8/4)^2 = 17.8$$

Hence, a minimum of 18 sample data would be required to estimate the mean height that is within 4 cm of the population mean and having 95% confidence level.

- **Illustration 3.3**

Everley's syndrome is a congenital disease that causes a reduction in concentration of blood sodium. Assuming blood sodium concentration to be normally distributed, what should be the sample size in estimating the mean blood sodium concentration among the patients in the population with 95% confidence and error margin of 5 mmol/l? Studies in the review suggest that the standard deviation of the blood sodium concentration ($S$) is 12 mmol/l.

- **Solution**

Here, we need to estimate the sample size for estimating mean blood sodium concentration among the patients in the population when

Error margin $(d) = 5\,\text{mmol}/1$

$\alpha = 0.05$

Standard deviation $(S) = 12\,\text{mmol}/1$

Let us assume $n = 20$, $t_{\alpha/2,v} = t_{0.05/2,19} = 2.093$. Since we know

$$n = \left(t_{\alpha/2,v}\right)^2 \times \left(\frac{S}{d}\right)^2$$
$$= 2.093^2 \times (12/5)^2 = 25.23$$

Now let us assume $n = 26$, $t_{0.05/2,25} = 2.060$

$$n = \left(t_{0.05/2,25}\right)^2 \times \left(\frac{S}{d}\right)^2$$
$$= 2.060^2 \times (12/5)^2 = 24.44$$

Hence, $n$ should be at least 25 to have 95% confidence estimate of mean sodium concentration that has 5 mmol/l margin of error.

- **Illustration 3.4**

A researcher is interested in estimating the mean serum indirect bilirubin level in 2-week-old infants. Similar studies suggest that standard deviation of the bilirubin level is 2.2 mg/dl. What should be the sample size in estimating mean bilirubin level which is within 1.2 mg/dl of the true mean with 99% confidence?

- **Solution**

Given $\alpha = 0.01$, $d = 1.2$ mg/dl, $S = 2.2$ mg/dl.

Let us assume that $n = 20$ then $t_{\alpha/2,v} = t_{.01/2,19} = 2.861$.

$$n = \left(t_{0.01/2,19}\right)^2 \times \left(\frac{S}{d}\right)^2$$
$$= 2.861^2 \times (2.2/1.2)^2 = 27.51$$

Since estimated sample size, 27.51 is higher than the assumed one, i.e. 20, let us assume, $n = 28$, $t_{0.01/2,27} = 2.771$.

$$n = \left(t_{0.01/2,27}\right)^2 \times \left(\frac{S}{d}\right)^2$$
$$= 2.771^2 \times (2.2/1.2)^2 = 25.8$$

Hence, sample size should be at least 26 to obtain 99% confidence estimate of mean serum indirect bilirubin within 1.2 mg/dl margin of error.

## Determining Sample Size in Estimating Population Proportion

In general, larger sample provides more precise estimate of unknown parameters. For example, if we wish to know the proportion of smokers, we would generally have a more precise estimate of this proportion if we examine a sample of 200 hundred subjects rather than 100. In certain situations, larger sample enhances minimal or even non-existent precision. This may be due to the presence of systematic errors or strong dependence in the data or due to the highly skewed data.

If $x$ is the number of observations having desired characteristics (say number of subjects out of the $n$ sampled subjects who are smokers), then $x/n$ is the estimator of the population proportion. If the observations are independent the estimator, $\hat{p} = x/n$ follows binomial distribution and is also the mean of the distribution. For sufficiently large $n$, the distribution of $\hat{p}$ is normally distributed with mean $p$ and variance $pq/n$ where $p$ is the proportion of the characteristics in the population and $n$ is the sample size. The maximum variance of this distribution would be $0.25/n$, which occurs when the population proportion is $p = 0.5$. Usually, $p$ is unknown; therefore, the maximum variance is often used for sample size estimation.

If $d$ is the margin of error on either side of the population proportion $p$, then 95% confidence limits of proportion shall be as follows:

$$\hat{p} - Z_{0.05/2}\sqrt{\frac{pq}{n}} \quad \text{to} \quad \hat{p} + Z_{0.05/2}\sqrt{\frac{pq}{n}}$$

From ◻ Fig. 3.2, we can see that

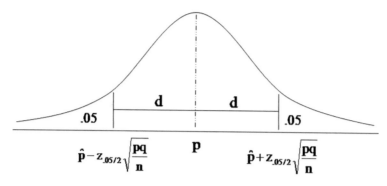

□ **Fig. 3.2** 95% confidence limits for the population proportion

$$2d = \left( \hat{p} + Z_{0.05/2}\sqrt{\frac{pq}{n}} \right) - \left( \hat{p} - Z_{0.05/2}\sqrt{\frac{pq}{n}} \right)$$

$$d = Z_{0.05/2}\sqrt{\frac{pq}{n}}$$

$$n = Z_{0.05/2}^2 \frac{pq}{d^2}$$

Since population proportion is usually unknown, hence, its estimate is obtained from the similar studies conducted earlier. Thus, using the estimate of $p$ as $\hat{p}$, the above equation becomes

$$n = Z_{0.05/2}^2 \frac{\hat{p}\hat{q}}{d^2} \tag{3.3}$$

For simplicity, let us take the approximate value of $Z_{0.05/2}$ as 2 and since $p$ is rarely known we shall take $p$ as 0.5 (assume maximum variance); then

$$n = 2^2 \frac{0.5 \times 0.5}{d^2} = \frac{1}{d^2} \tag{3.4}$$

Now, if we require an estimate of proportion within $\pm 10\%$ of the population proportion in a survey study, then $d$ will be equal to 0.1. Substituting this value in Eq. (3.4), $n$ will become 100. As a guideline, for different levels of precision ($d$), sample sizes may be calculated by the research scholar for 95% confidence level as shown in □ Table 3.1. Usually, in survey studies estimate of proportion is obtained with precision of 5% or less.

Thus, while estimating population proportion, the researcher needs to identify the following three things:

1. Degree of confidence in the estimation (95% or 99% or any other confidence level).
2. Degree of precision required in the estimation.
3. The magnitude of the population proportion ($p$) of the characteristics to be estimated in the population. It may be estimated from the review studies. Usually, it is unknown; therefore, maximum variance of the sample proportion is assumed and $p$ is taken as 0.5. Thus, $q$ also becomes 0.5.

| Precision (%) | $d$ | Sample size |
|---|---|---|
| 10 | 0.1 | 100 |
| 5 | 0.05 | 400 |
| 4 | 0.04 | 625 |
| 3 | 0.03 | 1111 |
| 2 | 0.02 | 2500 |
| 1 | 0.01 | 10000 |

◘ **Table 3.1** Sample size for different precisions at 95% confidence level

## Sample Size Determination for Estimating Proportion

- **Illustration 3.5**

In order to estimate the proportion of non-vegetarians in a city, what should be the sample size for 95% confidence interval for population proportion with a margin of error at 4%? In similar studies in estimating non-vegetarians in the city, proportion $\hat{p}$ was 0.3.

- **Solution**

Given that $Z_{0.05/2} = 1.96$, $d = 0.04$, and $\hat{p} = 0.3$.
   Since

$$n = Z_{0.05/2}^2 \frac{\hat{p}\hat{q}}{d^2}$$

$$n = 1.96^2 \frac{0.3 \times 0.7}{0.04^2} = 504.21$$

Thus, sample size of 504 is required to estimate the proportion of non-vegetarian with 95% confidence level and 4% margin of error.

- **Illustration 3.6**

In order to develop smoking policy, university authority is interested to estimate the proportion of non-smokers in the university. The proportion of non-smokers in similar studies was 0.7. How much should be the sample size to have an estimate within 3% of the true proportion with 95% confidence coefficient?

- **Solution**

If the sample size is larger than 30, the proportion of the characteristics can be considered to be normally distributed. Hence, by assuming normality the sample size n can be determined by using the formula

$$n = Z_{\alpha/2}^2 \frac{\hat{p}\hat{q}}{d^2}$$

Here $\alpha = 0.05$, $d = 0.03$, $\hat{p} = 0.7$, $\hat{q} = 0.3$, $Z_{\alpha/2} = Z_{0.05/2} = 1.96$

$$n = 1.96^2 \frac{0.7 \times 0.3}{0.03^2} = 896.37$$

A minimum of 897 sample data is required to have 95% confidence interval for the estimated proportion of non-smokers that will have 3% precision.

- **Illustration 3.7**

A physiotherapy centre provides its services to both male and female clients. It is decided to determine the proportion of female patients visiting the centre to develop a policy for deploying a proper proportion of manpower of therapists. Similar studies suggest that the proportion of female patients visiting such centres is 0.4. How large a sample of patients should be selected to estimate the proportion of female patients with 4% margin of error having 90% confidence level?

- **Solution**

Here, we know that $\alpha = 0.10$, $d = 0.04$, $\hat{p} = 0.4$, $\hat{q} = 0.6$, $Z_{\alpha/2} = Z_{0.10/2} = 1.645$

$$n = Z_{\alpha/2}^2 \frac{\hat{p}\hat{q}}{d^2}$$

$$= 1.645^2 \frac{0.4 \times 0.6}{0.04^2} = 405.9$$

A minimum of 406 sample data is required to have 90% confidence in the estimated proportion of the female patients visiting the therapy centre in the population that has 4% margin of error.

- **Illustration 3.8**

A survey reports that 15% of all inhabitants aged 16–22 years in a metro city drove under the influence of drugs or alcohol. A similar survey is planned for Delhi. They want a 95% confidence interval to have a 4% margin of error.
1. Find the estimated sample size if they expect the results similar to that of the metro city.
2. In the absence of the survey results by using the conservative formula based on $\hat{p} = 0.50$, what would be the new sample size?

- **Solution**

Given $\alpha = 0.05$, $d = 0.04$, $\hat{p} = 0.15$, $\hat{q} = 0.85$
1. We are given $Z_{\alpha/2} = Z_{0.05/2} = 1.96$

$$n = Z_{0.05/2}^2 \frac{\hat{p}\hat{q}}{d^2}$$

$$= 1.96^2 \frac{0.15 \times 0.85}{0.04^2} = 306.13$$

A sample size of 307 is required to have the same results in Delhi for estimating the proportion of inhabitants driving under the influence of drugs or alcohol with 95% confidence level and having 4% margin of error.

2. If $\hat{p}$ is not known and assumed to be 0.5, then $\hat{q}$ will also be 0.5.
$$Z_{\alpha/2} = Z_{0.05/2} = 1.96.$$

$$n = Z_{0.05/2}^2 \frac{\hat{p}\hat{q}}{d^2} = 1.96^2 \frac{0.50 \times 0.50}{0.04^2} = 600.25$$

Thus, if $\hat{p}$ is not known, then a sample of size 601 is required to have the estimate of proportion of inhabitants driving under the influence of drugs or alcohol with 95% confidence and having 4% error margin.

## Determining Sample Size in Estimating Difference Between Two Population Means

We have witnessed above the procedure of estimating the required sample size for a given confidence level and specified precision. Similar concept is used in estimating the sample size $n$, required from each of the two populations in order to estimate the difference between the two population means with specified precision. If the samples $x$ with size $n_1$ and $y$ with size $n_2$ are drawn from the two normal populations $N(\mu_1, \sigma_1^2)$ and $N(\mu_2, \sigma_2^2)$, respectively, then we know that

$$(\bar{x} - \bar{y}) \text{ follows } N\left(\mu_1 - \mu_2, \frac{\sigma_1^2}{n_1} + \frac{\sigma_2^2}{n_2}\right)$$

Assuming that the variances of the two populations are equal and sample sizes are same, the distribution of $\bar{x} - \bar{y}$ becomes

$$N\left(\mu_1 - \mu_2, \frac{2\sigma^2}{n}\right)$$

Thus, if $d$ is the maximum difference between the difference of means and its estimated value and $(1 - \alpha) \times 100\%$ is the confidence level, then

$$2d = \left((\bar{x} - \bar{y}) + Z_{\alpha/2}\sigma\sqrt{\frac{2}{n}}\right) - \left((\bar{x} - \bar{y}) - Z_{\alpha/2}\sigma\sqrt{\frac{2}{n}}\right)$$

or

$$2d = 2Z_{\alpha/2}\sigma\sqrt{\frac{2}{n}}$$

ꞌ

or

$$n = 2Z_{\alpha/2}^2 \frac{\sigma^2}{d^2} \tag{3.5}$$

If $\sigma^2$ is not known and is being estimated from the review studies, then instead of $Z_{\alpha/2}$ the value of $t_{\alpha/2,v}$ may be used and the sample size may be estimated by using the iteration process.

## Summary

In survey studies, determining sample size depends upon three factors; variability of the measure in the population, required precision of the estimate and confidence coefficient. Thus, sample size can be estimated by using the formula $n = (t_{\alpha/2,v})^2 \times (S/d)^2$ where $d$ is the desired precision ($\mu \pm d$), $S$ is the estimate of the variable's standard deviation in the population and $\alpha$ is the level of significance ($1 - \alpha$ is the confidence coefficient). On the other hand, sample size in estimating population proportion can be determined by using the formula $n = Z_{0.05/2}^2 (pq/d^2)$, where $p$ is the proportion of the characteristics in the population. Assuming maximum variance ($p = q$), the formula can be simplified as $1/d^2$ for estimating the sample size if the population is very large.

- Exercises

1. How does hypothesis testing relate to sample size determination?
2. When do you reject the null hypothesis? Should the null hypothesis be accepted if the test statistic warrants so?
3. In estimating population mean if the tolerable amount of error is $\pm d$ and $\sigma$ is the population standard deviation, derive the formula for determining the sample size.
4. What are the various considerations in deciding sample size in survey studies?
5. In estimating population proportion derive the formula in determining the sample size by assuming maximum variance of the characteristics in the population.
6. In determining the environment of the organization, the company plans to investigate whether employees are happy or not. The company is willing to accept a margin of error of 4% but wants 95% confidence. How many randomly selected employees the study should include to estimate proportion of employees who are happy?
7. A psephologist is interested to know as to what percentage of voters will vote in the coming assembly election in the city. What sample size he should take to ensure that the margin of error is less than 3% assuming 99% level of confidence?
8. A researcher wishes to estimate the proportion of students who practice meditation. How many subjects should be randomly selected if he wishes the estimate to be within 0.02 with 95% confidence in the following two situations?
   (a)  He uses the information about the previous estimate of 0.35.
   (b)  He does not have any prior knowledge about the estimate.
9. The standard deviation of weight of the students in a university is known to be 3 kg. What should be the size of the sample if the investigator wishes to estimate population mean weight with 95% confidence with sampling error of 0.5 kg or less?

10. Let us suppose you wish to estimate the height of women with a margin of error 10 cm, and a preliminary study indicates that the women weights are normally distributed with a standard deviation of 55 cm. What sample size you should take in the study if the required confidence level is 99%?

**3**

■ **Answer**

6. $n = 600$ (Hint: take $p = 0.5$)
7. $n = 1849$ (Hint: take $p = 0.5$)
8. (a)    $n = 2185$
   (b)    $n = 2401$
9. $n = 138$
10. $n = 201$

## Bibliography

Abramson, J. J., & Abramson, Z. H. (1999). *Survey methods in community medicine: Epidemiological research, programme evaluation, clinical trials* (5th ed.). London: Churchill Livingstone/Elsevier Health Sciences. ISBN 0-443-06163-7.

Bartlett, J. E., II., Kotrlik, J. W., & Higgins, C. (2001). Organizational research: Determining appropriate sample size for survey research (PDF). *Information Technology, Learning, and Performance Journal, 19*(1), 43–50.

Beam, G. (2012). *The problem with survey research* (p. xv). New Brunswick, NJ: Transaction.

Francis, J. J., Johnston, M., Robertson, C., Glidewell, L., Entwistle, V., Eccles, M. P., et al. (2010). What is an adequate sample size? Operationalising data saturation for theory-based interview studies. *Psychology and Health, 25*, 1229–1245. ► https://doi.org/10.1080/08870440903194015.

Galvin, R. (2015). How many interviews are enough? Do qualitative interviews in building energy consumption research produce reliable knowledge? *Journal of Building Engineering, 1*, 2–12.

Guest, G., Bunce, A., & Johnson, L. (2006). How many interviews are enough? An experiment with data saturation and variability. *Field Methods, 18*, 59–82. ► https://doi.org/10.1177/1525822X05279903.

Jung, S. H., Kang, S. H., & Ahn, C. (2001). Sample size calculations for clustered binary data. *Statistics in Medicine, 20*(1971–1982), 2001.

Kish, L. (1965). *Survey sampling*. Wiley. ISBN 0-471-48900-X.

Kish, L. (1965). *Survey sampling*. Wiley. ISBN 978-0-471-48900-9.

Leung, W.-C. (2001). Conducting a survey, in student BMJ, (British Medical Journal, Student Edition), May 2001.

Onwuegbuzie, A. J., & Leech, N. L. (2007). A call for qualitative power analyses. *Quality & Quantity, 41*, 105–121. ► https://doi.org/10.1007/s11135-005-1098-1.

Ornstein, M. D. (1998). Survey research. *Current Sociology, 46*(4), iii–136.

Sandelowski, M. (1995). Sample size in qualitative research. *Research in Nursing & Health, 18*, 179–183.

Smith, S. (2013, April 8). *Determining sample size: How to ensure you get the correct sample size*. Qualtrics. Retrieved November 15, 2016.

# Understanding Concepts in Estimating Sample Size in Hypothesis Testing Experiments

© Springer Nature Singapore Pte Ltd. 2020
J. P. Verma and P. Verma, *Determining Sample Size and Power in Research Studies*,
https://doi.org/10.1007/978-981-15-5204-5_4

**4**

◘ **Learning Objectives**

After going through this chapter, the readers should be able to

– Explain how decision about null hypothesis is affected with the change in sample size for the same boundary conditions.
– Fix the boundary conditions in hypothesis testing experiments.
– Understand the procedure in computing sample size and power in one- and two-sample testing.
– Compute sample size and power in hypothesis testing experiments related to mean and proportion.

## Introduction

In undertaking any research study, we attempt to focus on answering some of the research issues for enhancing, updating and advancing existing knowledge. These research issues can be answered by using either inductive or deductive logic approaches. In inductive approach, researcher investigates observations and analyses different trends and occurrences and if certain trends are important then a hypothesis is framed and tested, and thereafter theory is built. For instance, in a survey study after collecting data, one may like to investigate the buying behaviour of the people in different socioeconomic groups. In such studies, nothing is fixed a priori. It is only after collecting the data and visualizing some trends that different hypotheses are formed and tested. If hypothesis testing confirms some kind of relationship, we may build a theory that the buying behaviour is affected by socioeconomic status. Consider a situation in which a large number of patients start visiting doctors in the last few weeks; the researcher may develop a theory that existing weather is not conducive for the inhabitants or there is some outbreak of disease(s) in the neighborhood. Similarly, if majority of the medal winners in sports are being produced by a particular sports university, one may develop a theory that the training program of the university is superior to all other similar universities. Studies based on inductive logics are common in social sciences. Thus, in inductive research, theories are built.

On the other hand, studies based on deductive logic start with the assumption that the theory exists and the hypothesis is tested to validate it in a particular situation. Such studies are common in natural sciences. For instance, we have a theory that aerobic exercise improves cardiorespiratory endurance; now a research scientist may plan a study in which he may like to vary the exercise intensity (low and medium) in a controlled environment to observe whether exercise with medium intensity improves cardiorespiratory endurance more than that of low intensity.

If you compare these two types of research, inductive approach is more flexible in nature than the deductive one. Nothing is fixed in advance, and it allows researchers to explore different ideas before determining which one correlates to a hypothesis. The conclusion drawn in the inductive research lacks cause and effect relationship in comparison to the deductive research. It is because of the fact that in inductive research the researcher is not in a position to manipulate independent variable, hence, the study is not undertaken in a controlled environment, whereas, in deductive research, the researcher controlling for other factors, varies only the independent variable of interest to see its effect on another variable. Since, deductive research is conducted in a controlled setting, it explains more cause and effect relationship while requiring fewer subjects. In this chapter, we will understand the concept in determining the sample size in hypothesis testing experiment.

## Importance of Sample Size in Experimental Studies

Generally, researchers believe that larger the sample, more efficient is the finding. Let us investigate this fact in experimental studies. The purpose of any experimental research is to investigate the cause and effect relationship between an intervention and some variable of interest. In doing so, the researcher's interest is to see whether the intervention is effective, which is tested by means of p-value. If the p-value is less than 0.05, then the researcher is 95% confident that the effect exists. In fact, significance has got no meaning until and unless the size of the effect is known. For instance, if a drug manufacturing company claims that they are 95% confident that their newly developed drug is effective in reducing diastolic pressure of the hypertensive patient, it simply means that out of the patients who consume drug, 95% will reduce their diastolic pressure. This finding has been obtained because p-value associated with t-test was less than 0.05. But how much reduction will take place is not revealed. If the drug reduces only 2 mm Hg in the patient's diastolic pressure, the result is meaningless. Thus, until the researcher reports p-value along with achieved effect, the result is meaningless. The researcher needs to decide the meaningful effect size before starting an experiment because sample size depends upon the size of the effect one requires in the study. The effect size can be considered as the minimum detectable difference in testing the hypothesis. In larger sample, even small effect will reject the null hypothesis whereas in smaller sample, null hypothesis may not be rejected even for large effect if all other conditions are same. Thus, it is important to find the appropriate sample size for detecting desired effect of the intervention. Let us understand this fact by means of an illustration.

Let us assume that in a random sample of 16 hypertensive female the mean body mass index (BMI) is 30 kg/m². It is known that the mean and standard deviation of the BMI of the hypertensive female are 26 and 4 kg/m², respectively. If the investigator wishes to test at 5% level, whether these 16 observations have come from the population with a mean of 26 kg/m², the z-test can be applied as the standard deviation of the hypothesized population is known. Here, the null hypothesis $H_0: \mu = 26$ is tested against the alternative hypothesis $H_a: \mu \neq 26$. The value of z can be computed by the following formula:

$$z = \frac{\bar{x} - \mu}{\sigma/\sqrt{n}}$$

Now the null hypothesis will be rejected if the value z is greater than 1.96 or $(\bar{x} - \mu)$ is greater than $1.96 \times \frac{4}{\sqrt{n}}$. In other words, the null hypothesis will be rejected at 5% level if the below-mentioned equation is true.

$$(\bar{x} - \mu) > \frac{7.84}{\sqrt{n}}$$

| ☐ **Table 4.1** Threshold difference with varying sample size for rejecting null hypothesis | Sample size ($n$) | Threshold difference $(\bar{x} - \mu)$ |
|---|---|---|
| | 16 | 1.96 |
| | 64 | 0.98 |
| | 144 | 0.65 |
| | 256 | 0.49 |

**4**

This indicates that the rejection of null hypothesis depends upon the sample size. The threshold difference $(\bar{x} - \mu)$ for different sample sizes in the same experiment at 5% level can be computed as shown in ☐ Table 4.1.

☐ Table 4.1 reveals that if the sample size is 16 then a difference between sample and population mean greater than 1.96 will reject the null hypothesis, whereas if the sample is increased to 256 then even the difference greater than 0.49 will make the null hypothesis rejected provided the level of significance is same. Thus, one needs to decide as to how much effect is required to be tested besides other boundary conditions in the experiment; accordingly, sample size needs to be decided.

## Sample Size on the Basis of Power

In hypothesis testing experiments, sample size is determined on the basis of power required to detect a specific effect in the study. We have earlier seen in ▶ Chap. 2 that the power of a test is the probability in rejecting a null hypothesis when alternative hypothesis is true. In other words, if the claim is good it should be accepted by the test. While discussing the relationship between Type I and Type II errors, we have seen that if power increases, the Type I error also increases. Hence, the researcher should judiciously decide the power $(1 - \beta)$ and level of significance ($\alpha$) in their study before performing the hypothesis testing experiment. Besides Type I error, power is also affected by the sample size, variability of the population, minimum detectable difference and the type of test (one tailed or two tailed). Generally, power in any hypothesis testing experiment must be at least 0.8 or more. Just like power, sample size is also affected by the power of the test, Type I error, population variability, minimum detectable difference and the type of test (one tailed or two tailed). We shall now understand the mechanism in estimating sample size for specific power to detect a particular amount of effect in different hypothesis testing experiments.

## One-Sample Testing of Mean

## Determining Sample Size

We have seen above that the sample size in a hypothesis testing experiment depends upon the amount of effect to be detected with some specified power. Thus, in one-sample test to decide the sample size, one should first decide the effect size to be detected and power in the study. Further, estimating sample size ($n$) also depends upon the population variance $\sigma^2$. Usually, the population variance is unknown and is estimated by $S^2$ as reported in the similar studies conducted previously. Alternatively, it may be estimated by conducting a pilot study.

In performing $t$-test, we may specify the probability of committing Type I error as $\alpha$, the probability of Type II error as $\beta$ and minimum detectable difference between means under $H_0$ and $H_1$ as $d$. Thus, to test the hypothesis using $t$-test at the significance level $\alpha$ with power $(1 - \beta)$, the estimate of sample size required to detect the difference $d$ is given by

$$n = \frac{S^2}{d^2} \left(t_{\alpha,v} + t_{\beta,v}\right)^2$$

Here $t_{\alpha,v}$ represents value of $t$ in the distribution for one-tailed test with significance level $\alpha$ and $v$ degrees of freedom. For two-tailed test, $t_{\alpha,v}$ should be replaced by $t_{\alpha/2,v}$. The degrees of freedom $v$ can be obtained only if the sample size n is known. Since n cannot be calculated directly, it will be known by iteration. This is shown in Illustration 4.1.

In case the population variance $\sigma^2$ is known, then sample size $n$ can be estimated by using the normal distribution and the formula will become

$$n = \frac{\sigma^2}{d^2} \left(Z_\alpha + Z_\beta\right)^2$$

The value of $Z_\alpha$ is the value of $Z$ in the distribution for one-tailed test with significance level $\alpha$. For two-tailed test, $Z_\alpha$ should be replaced by $Z_{\alpha/2}$.

---

**Derivation of Formula for Estimating Sample Size in One-Sample Testing**

If population variance $\sigma$ is known, then the null hypothesis $H_0:\mu=\mu_0$ can be tested against the alternative hypothesis $H_1 : \mu > \mu_0$ by using the following z statistic:

$$Z = \frac{\overline{X} - \mu}{\sigma/\sqrt{n}}$$

If level of significance is $\alpha$, then $H_0$ is rejected if $Z > C$ (◪ Fig. 4.1). The value of $C$ under null and alternative hypotheses can be obtained as follows:

Under $H_0$, $\quad C = \mu_0 + z_\alpha \dfrac{\sigma}{\sqrt{n}}$ $\qquad\qquad$ (4.1)

Under $H_1$, $\quad C = \mu_1 - z_\beta \dfrac{\sigma}{\sqrt{n}}$ $\qquad\qquad$ (4.2)

Equating (4.1) and (4.2)

**4**

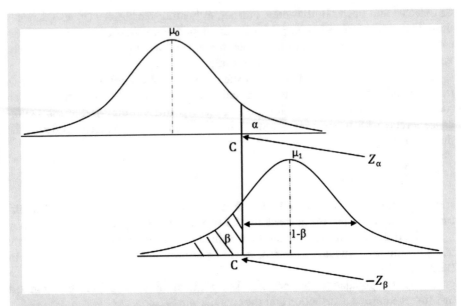

☐ **Fig. 4.1**   Showing areas under the null and alternative hypotheses for a given $\alpha$ and $\beta$

$$\mu_0 + Z_\alpha \frac{\sigma}{\sqrt{n}} = \mu_1 - Z_\beta \frac{\sigma}{\sqrt{n}}$$

$$\Rightarrow \quad \mu_1 - \mu_0 = \frac{\sigma}{\sqrt{n}} (Z_\alpha + Z_\beta) \quad [\text{Taking } d = \mu_1 - \mu_0]$$

$$\Rightarrow \quad n = \frac{\sigma^2}{d^2} (z_\alpha + z_\beta)^2 \tag{4.3}$$

If population standard deviation ($\sigma$) is not known and is estimated by $S$, then the sample size in that case may be estimated by using the formula

$$n = \frac{S^2}{d^2} (t_{\alpha,v} + t_{\beta,v})^2 \tag{4.4}$$

**Remark** For two-tailed test, $t_{\alpha,v}$ should be replaced by $t_{\alpha/2,v}$.

## Estimation of Sample Size in One-Sample Test

■  **Illustration 4.1**

How large a sample is required to reject the null hypothesis $H_0\!:\!\mu = 37$ against the alternative hypothesis $H_1\!:\!\mu > 37$, where $\mu$ is the population mean of VO2max (a sign of cardiorespiratory endurance)? We wish to test the null hypothesis at 0.05 level of significance with 80% chances of detecting the difference between the means under

$H_1$ and $H_0$ by at least 1.2 units ($\mu_1 - \mu_0 > 1.2$). Population variance has been estimated by the reported studies as $S^2 = 4$ ml kg$^{-1}$ min$^{-1}$.

- **Solution**

Let us assume that the estimated sample size ($n$) is 20 in testing the above hypothesis.

Then $v$ (degrees of freedom) $= n - 1 = 19$, for one-tailed test

$$t_{\alpha,v} = t_{0.05,19} = 1.729$$

$$t_{\beta,v} = t_{0.20,19} = 0.861, \quad (\text{Since } \beta = 1 - 0.80 = 0.20)$$

$$d = 1.2$$

$$S^2 = 4$$

Substituting these values in Eq. (4.4)

$$n = \frac{4}{1.2^2}(1.729 + 0.861)^2 = 18.63$$

Since estimated sample size is 18.63 which is less than the assumed 20, hence, let the new estimate of the sample size $n = 19$. In that case $v = 18$, $t_{\alpha,v} = t_{0.05,18} = 1.734$, $t_{\beta,v} = t_{0.20,18} = 0.862$ (Since $\beta = 1 - 0.80 = 0.20$). The new estimate of sample size will be

$$n = \frac{4}{1.2^2}(1.734 + 0.862)^2 = 18.72$$

Thus, it may be concluded that a sample of 19 data may be taken from this population to test the above hypothesis for the significance level 0.05, power 0.8 and minimum detectable difference 1.2.

## Estimation of Minimum Detectable Difference

By reorganizing Eq. 4.4, we can find the minimum detectable difference between the means under $H_1$ and $H_0$ by the t-test with $(1 - \beta)$ power, at $\alpha$ level of significance, using a sample of size $n$.

$$d = \frac{S}{\sqrt{n}}\left(t_{\alpha,v} + t_{\beta,v}\right) \tag{4.5}$$

where $t_{\alpha,v}$ represents value of $t$ in one-tailed test and for two-tailed test it should be replaced by $t_{\alpha/2,v}$.

- **Illustration 4.2**

In illustration 4.1 where one-tailed test was used, what is the minimum detectable difference (between the means under $H_1$ and $H_0$) that can be detected 80% of the time using a sample of 24 data at 5% level?

**4**

- **Solution**

Using Eq. (4.5)

$$d = \frac{2}{\sqrt{24}}(t_{0.05,23} + t_{0.20,23})$$

From Table A2 in Appendix, $t_{0.05,23} = 1.714$, $t_{0.20,23} = 0.857$.
Thus,

$$d = \frac{2}{\sqrt{24}}(1.714 + 0.857) = 1.05$$

## Estimation of Power in One-Sample $t$-Test

Equation 4.4 can be used to estimate the power in one-sample test for a given sample size $(n)$, $\alpha$, type of test (one or two tailed) and population variability. After rearranging the equation

$$t_{\beta,v} = \frac{d \times \sqrt{n}}{s} - t_{\alpha,v} \qquad (4.6)$$

- **Illustration 4.3**

In illustration 4.1 if the sample size is 24, what would be the power of the one-tailed test for a given $\alpha = 0.05$, $d$ (minimum detectable difference) $= 1.2$ and $S = 2$?

- **Solution**

Substituting values in Eq. 4.6

$$t_{\beta,23} = \frac{1.2 \times \sqrt{24}}{2} - t_{0.05,23}$$

If we assume that the population variance is known, then value of $t_{0.05,23}$ can be determined by $Z_{0.05}$ ($=1.645$ for one-tailed test) and power can be determined by using the $z$-distribution instead of t as shown in ◘ Fig. 4.2.
    Thus,

$$Z_\beta = \frac{1.2 \times \sqrt{24}}{2} - 1.645 = 1.29$$

From the normal area Table A1 in Appendix, it can be seen that the $\beta = 0.0985$. Thus, the power of the test would be 90.15%.

## Testing Difference Between Two Means

## Determining Sample Size in Two-Sample $t$-Test

Let us see how to determine the sample size in two-sample $t$-test for desired power in detecting a given difference between two population means. Suppose we wish to test the null hypothesis $H_0$: $\mu_1 - \mu_2 = 0$ against the alternative hypothesis $H_1$: $\mu_1 - \mu_2 > 0$.

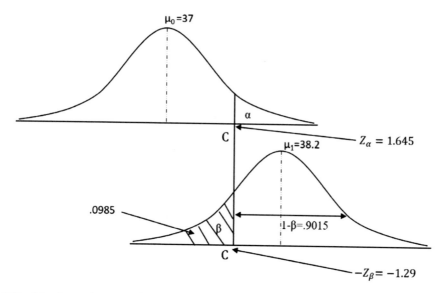

□ **Fig. 4.2** Computing power for given sample size

In order to test the significance difference between the two group means at $\alpha$ level of significance, with a $1 - \beta$ power to detect a true difference between population means as small as $d(\mu_1 - \mu_2)$, the estimated sample size can be obtained by the following formula:

$$n = \frac{2\sigma^2}{d^2}\left(Z_\alpha + Z_\beta\right)^2$$

where $\sigma^2$ is the population variance. In case population variance is unknown and is being estimated by $S^2$, then the sample size is estimated by using the following formula:

$$n = \frac{2S^2}{d^2}\left(t_{\alpha,v} + t_{\beta,v}\right)^2$$

**Derivation of Formula for Estimating Sample Size in Two-Sample t-Test**
Suppose we wish to test the null hypothesis $H_0: \mu_1 - \mu_2 = 0$ against the alternative hypothesis $H_1: \mu_1 - \mu_2 > 0$ at $\alpha$ significance level. We know that if the two samples of the same size $(n)$ are drawn from the two normal populations having the same variance $\sigma^2$, then

$(\bar{x}_1 - \bar{x}_2)$ follows $N\left((\mu_1 - \mu_2), \dfrac{2\sigma^2}{n}\right)$

**4**

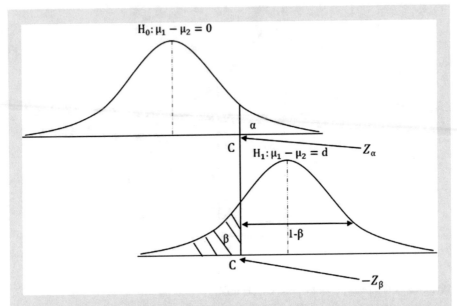

◻ **Fig. 4.3** Showing areas under null and alternative hypothesis for a given $\alpha$ and $\beta$

To test the null hypothesis, we compute

$$Z = \frac{\bar{x}_1 - \bar{x}_2}{\sigma}\sqrt{\frac{n}{2}}$$

and we reject $H_0$ if $Z > C$ under $H_0$ (◻ Fig. 4.3). The value of $C$ under the null and alternative hypotheses can be obtained as follows:

Under $H_0$,     $C = 0 + Z_\alpha \times \sigma\sqrt{\dfrac{2}{n}}$     (4.7)

Under $H_1$     $C = d - Z_\beta \times \sigma\sqrt{\dfrac{2}{n}}$     (4.8)

Equating (4.7) and (4.8)

$$0 + Z_\alpha \sigma\sqrt{\frac{2}{n}} = d - Z_\beta \sigma\sqrt{\frac{2}{n}}$$

$$\Rightarrow \qquad d = \sigma\sqrt{\frac{2}{n}}(Z_\alpha + Z_\beta)$$

or     $n = 2\dfrac{\sigma^2}{d^2}(Z_\alpha + Z_\beta)^2$     (4.9)

Since population standard deviation is usually not known and $\sigma$ is estimated by S, the sample size in such situations may be estimated by iteration method using the following formula:

$$n = \frac{2 \times S^2}{d^2}\left(t_{\alpha,v} + t_{\beta,v}\right)^2 \qquad (4.10)$$

**Remark** For two-tailed test, $t_{\alpha,v}$ should be replaced by $t_{\alpha/2,v}$.

■ **Illustration 4.4**

In a study, the researcher desires to test whether the difference between the mean WBC counts of persons using two different drugs is significant. He wishes to test the hypothesis at 0.05 significance level with a 90% chance of detecting a true difference between the population means being as small as 40 counts. Estimate of the population variance as suggested by the earlier studies is $S^2 = 1700$ count. What should be his sample size?

■ **Solution**

Given that $\alpha = 0.05$, $\beta = 1-0.9 = 0.10$, $d = 40$, $S^2 = 1700$. For this two-tail test, let us assume that the estimated sample size is $n = 20$. To determine the exact sample size, we shall use the Formula 4.10. Let us first find the values of $t_{\alpha/2,v}$ and $t_{\beta,v}$ from the table. Here $v = 2(n-1) = 2(20-1) = 38$.

From Table A2 in Appendix, we get $t_{\alpha/2,v} = t_{0.05/2,38} = 2.023$ and $t_{\beta,v} = t_{0.10,38} = 1.304$

$$n = \frac{2 \times s^2}{d^2}\left(t_{\alpha/2,v}, + t_{\beta,v}\right)^2$$

or

$$n = \frac{2 \times s^2}{d^2}\left(t_{0.05/2,38} + t_{0.1,38}\right)^2$$
$$= \frac{2 \times 1700}{1600}(2.023 + 1.304)^2 = 23.52$$

Since the estimated sample size is higher than the assumed 20, hence, let us take the new estimate of the sample size as 24. Then, $v = 24+24-2 = 46$, $t_{\alpha/2,v} = t_{0.05/2,46} = 2.01$ and $t_{\beta,v} = t_{0.1,46} = 1.30$. Substituting these values in the formula, we get

$$n = \frac{2 \times s^2}{d^2}\left(t_{0.05/2,46} + t_{0.1,46}\right)^2$$
$$= \frac{2 \times 1700}{1600}(2.01 + 1.30)^2 = 23.35$$

Hence, it is concluded that each of the two samples should have at least 24 data.

If in any one sample a smaller number of data is available say $n_1 = 21$, then the size of the other sample should be increased in order to have the same power in testing. The size of the second sample $n_2$ can be estimated as

$$n_2 = \frac{n \times n_1}{2n_1 - n} = \frac{24 \times 21}{2 \times 21 - 24} = 28$$

Thus, the second sample must have 28 data to retain the same power in the testing.

**4**

## Estimation of Power in Two-Sample *t*-Test

In two-sample *t*-test, with $n$ sample in each group and $\alpha$ level of significance, power can be estimated by manipulating Eq. 4.10. By rearranging the equation, we get

$$t_{\beta,v} = \frac{d}{\sqrt{2S^2/n}} - t_{\alpha,v} \tag{4.11}$$

■ Illustration 4.5

In two-tailed test of illustration 4.4, what would be the probability of detecting a difference of 40 counts between mean WBC counts of persons using the two drugs if $n_1 = n_2 = 22$ and $\alpha = 0.05$?

■ Solution

Here, we are given $d = 40$, $S^2 = 1700$. For $n_1 = n_2 = 22$, $t_{0.05/2,42} = 2.019$ (from Table A2 in Appendix)

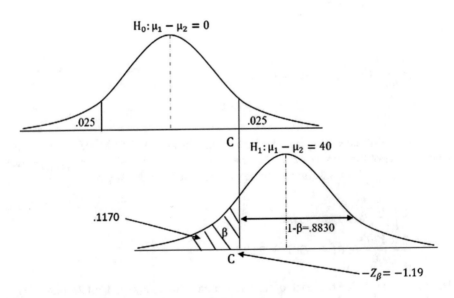

□ **Fig. 4.4**   Computing power for given sample size

$$t_{\beta,v} = \frac{d}{\sqrt{2S^2/n}} - t_{0.05/2,42}$$

$$= \frac{40}{\sqrt{2 \times 1700/22}} - 2.019 = 1.19$$

By normal approximation, we can estimate $\beta$ by $P(Z \geq 1.19) = 0.1170$ as shown in ☐ Fig. 4.4. Thus, the power of this two-sample $t$-test is 0.8830 or 88.3%.

## Summary

In hypothesis testing experiments, determining sample size is based on the concept of power. Sample size is estimated for a desired power by deciding $\alpha$, minimum detectable difference $d$, type of test (one/two-tailed test) and estimating the population standard deviation $\sigma$. Thus, in one-sample testing of mean, the sample size can be determined by using the formula $n = (s^2/d^2)(t_{\alpha,v} + t_{\beta,v})^2$. By manipulating this formula, minimum detectable difference and power can be estimated in the study. In two-sample testing for comparing group means, sample size can be determined by using the formula $n = (2s^2/d^2)(t_{\alpha,v} + t_{\beta,v})^2$. The value of $t_{\alpha,v}$ is used in case of one-tailed test, whereas $t_{\alpha/2,v}$ is used if the two-tailed hypothesis is tested.

- **Exercises**
  1. Differentiate between researches based on inductive and deductive logics. Whether researchers frame hypotheses in both types of researches, if so at what stage?
  2. What is $p$-value? Is it sufficient to reject the null hypothesis only on the basis of $p$-value?
  3. What do you mean by the effect size and how it is different from $p$-value? Why Cohen's guideline is different for normal and biological experiments.
  4. How the change in sample size affects the decision in hypothesis testing experiments if other things are constant? Show by means of an example.
  5. What is power and how it is calculated in one-sample $t$-test?
  6. Why power increases with the increase in sample size and what is the desired level of power one should have?
  7. Derive the formula for estimating sample size in one-sample testing of mean.
  8. What are the roles of effect size and power in a hypothesis testing experiment? How they affect each other?
  9. Differentiate unstandardized and standardized effect size. If review studies are not available to suggest different parameters to compute the effect size, how will you decide it for your experiment?
  10. When one should use unstandardized effect size over standardized effect size?
  11. How will you estimate the sample size in comparing means of two groups? Derive the formula.
  12. After introducing new policy, a company desires to investigate whether mean satisfaction score of the employees is more than 42. Satisfaction level is measured

on a scale on 10–50, where larger score indicates higher satisfaction. It is known that the satisfaction score is normally distributed with a standard deviation of 6. What sample size should be taken to detect the difference in means by 2 score with 90% power at the significance level of 0.05?

13. A researcher wishes to investigate whether performance of the students in private school is better in comparison to that of the students in government school in a particular age category. He wishes to detect the difference in their performance by three scores with 80% power at significance level 0.05. If variance of the performance score is known to be 81, what should be the sample size?

**4**

■ **Answer**

12. $n = 95$.

13. In each group, sample size should be equal to 141.

# Bibliography

Aberson CL (2010) *Applied power analysis for the behavioral science.* ISBN 1-84872-835-2.

Betz, M. A., & Gabriel, K. R. (1978). Type IV errors and analysis of simple effects. *Journal of Educational Statistics, 3*(2), 121–144.

Chow, S. C., Shao, J., Wang, H. (2008). *Sample size calculations in clinical research.* Chapman & Hall/CRC.

Ellis, P. (2010). *The essential guide to effect sizes: Statistical power, meta-analysis, and the interpretation of research results* (p. 52). Cambridge University Press. ISBN 978-0521142465.

Erdfelder, E., Faul, F., & Buchner, A. (2005). Power analysis for categorical methods. In B. S. Everitt & D. C. Howell (Eds.), *Encyclopedia of statistics in behavioral science* (pp. 1565–1570). Chichester: Wiley.

Hern, R. P. (2001). Sample size tables for exact single–Stage phase II designs. *Statistics in Medicine, 20,* 859–866.

Julious, S. A. (2004). Tutorial in biostatistics: Sample size for clinical trials. *Statistics in Medicine, 23,* 1921–1986.

Julious, S. A. (2009). *Sizes for clinical trials.* Chapman & Hall/CRC.

Julious, S. A., & Campbell, M. J. (2010). Tutorial in biostatistics: Sample size for parallel group clinical trials with binary data. *Statistics in Medicine, 31,* 2904–2936.

Krzywinski, M., & Altman, N. (2013). Points of significance: Power and sample size. *Nature Methods, 10,* 1139–1140.

Lenth, R. V. (2001). Some practical guidelines for effective sample size determination. *American Statistician, 55*(3), 187–193.

Muller, K. E., & Benignus, V. A. (1992). Increasing scientific power with statistical power. *Neurotoxicology and Teratology, 14,* 211–219.

Muller, K. E., Lavange, L. M., Ramey, S. L., & Ramey, C. T. (1992). Power calculations for general linear multivariate models including repeated measures applications. *Journal of American Statistical Association, 87*(420), 1209–1226.

Sedlmeier, P., & Gigerenzer, G. (1989). Do studies of statistical power have an effect on the power of studies? *Psychological Bulletin, 105,* 309–316.

Thomas, L. (1997). Retrospective power analysis. *Conservation Biology, 11*(1), 276–280.

Tsang, R., Colley, L., Lynd, L. D. (2009). Inadequate statistical power to detect clinically significant differences in adverse event rates in randomized controlled trials. *Journal of Clinical Epidemiology, 62*(6), 609–616. PMID 19013761. ▶ https://doi.org/10.1016/j.jclinepi.2008.08.005.

Wittes, J. (1984). Sample size calculations for randomized clinical trials. *Epidemiologic Reviews, 24*(1), 39–53.

# Use of G*Power Software

© Springer Nature Singapore Pte Ltd. 2020
J. P. Verma and P. Verma, *Determining Sample Size and Power in Research Studies*,
https://doi.org/10.1007/978-981-15-5204-5_5

■ Learning Objectives

After going through this chapter, the readers should be able to

— Understand the procedure in the installation of G*Power software.

— Describe downloading G*Power software in Windows as well as Mac operating systems.

## Introduction

G*Power is a general power analysis programme, which is meant for determining the sample size and analysing power in research studies. The concept and the procedures involved in determining the sample size in survey studies and hypothesis testing experiments have been discussed earlier in ▶ Chaps. 3 and 4. We will now discuss the procedure of downloading and installing G*Power software. In the following two chapters (▶ Chaps. 6 and 7), we will discuss its usage to carry out sample size determination and power analysis in hypothesis testing experiments. The G*Power is a free software which can be easily downloaded from the internet. In this chapter, we shall explain the procedure of downloading this software.

## Procedure of Installing G*Power 3.1

By following the steps below, G*Power software can be installed on your computer.

Step 1: Click on this link or paste it in your browser ▶ http://www.gpower.hhu.de

Step 2: After clicking on the link mentioned above, you will get the screen as shown in ◘ Fig. 5.1. On this screen after scrolling down you will get two options, one for Windows and another for Mac. As per your computer's operating system, select the appropriate link. Click on it to download the Setup file of the software.

Step 3: Downloading Setup file of the software will take a few minutes. Once the software is downloaded, a folder with the name G*Power_3.1.9.2 will appear on the screen as shown in ◘ Fig. 5.2.

Step 4: Click the folder G*Power_3.1.9.2 to see its files. One of the files is GPower-Setup. Click this Setup file to install the software.

Step 5: The computer will ask your permission for running the software. Do not worry; ignore this security option and click on *Run* (◘ Fig. 5.3).

Step 6: Click on *Next* for allowing Setup Wizard to install the software.

Step 7: Use default option in Folder window of the screen as shown in ◘ Fig. 5.4 so that the computer can install the software in the G*Power folder of the Program Files.

Step 8: Click on *Next* to start installation.

Step 9: During software installation, the screen shown in ◘ Fig. 5.5 will appear which shows the progress in installation. It will take a few minutes.

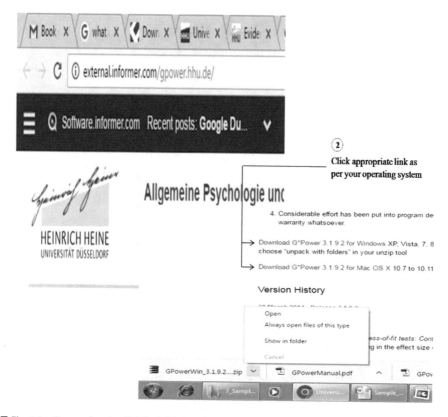

**Fig. 5.1** Screen showing link for initiating the process in downloading G*Power software

Step 10: After the installation of G*Power software, follow the sequence of commands shown in ☐ Fig. 5.6 to run the software in determining the sample size and analysing power in different situations as discussed in ► Chap. 6.

## Summary

The freeware G*Power software can be installed using the link ► http://www.gpower. hhu.de on the system. This software is available for Windows as well as for MAC operating system. After choosing the operating system, the G*Power software can be downloaded which will be shown in a folder named G*Power. Click the file named GPowerSetup to install the software in the system. Once the software is installed, it is ready for determining the sample size as well as power in different applications.

5

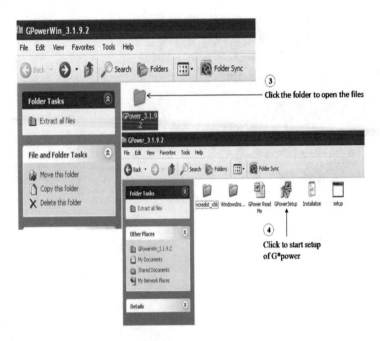

**◘ Fig. 5.2**   Screen showing downloaded folder of G*Power and its files

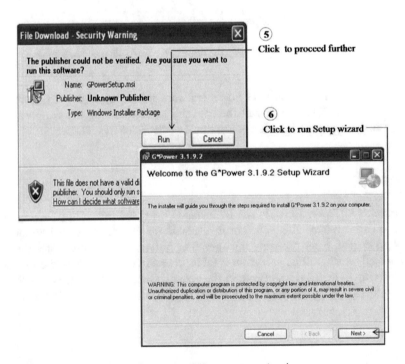

**◘ Fig. 5.3**   Screen showing option for running G*Power setup wizard

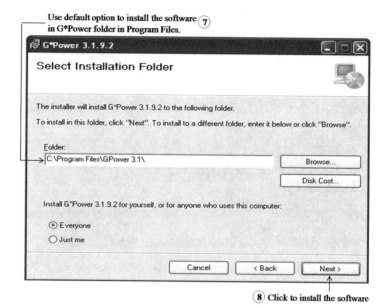

■ **Fig. 5.4** Screen for deciding the location where G*Power software needs to be installed

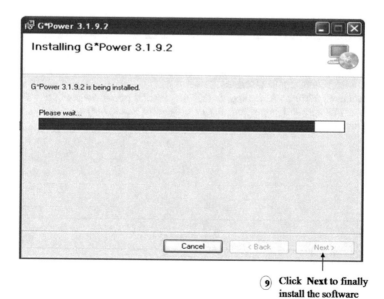

■ **Fig. 5.5** Screen showing the progress of installing the G*Power software

5

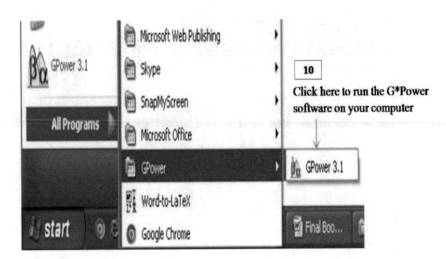

**□ Fig. 5.6**   Sequence of commands for running the G*Power software on computer

- **Exercises**

Q1. What is G-power in statistics?

Q2. What are the usual parameters required for computing sample size?

## Bibliography

Faul, F., Erdfelder, E., Lang, A., & Buchner, A. (2007). G*Power 3: A flexible statistical power analysis program for the social, behavioral, and biomedical sciences. *Behavior Research Methods, 39*(2), 175–191. ▶ https://doi.org/10.3758/bf03193146.

Faul, F., Erdfelder, E., Buchner, A., & Lang, A. (2009). Statistical power analyses using G*Power 3.1: Tests for correlation and regression analyses. *Behavior Research Methods, 41*(4), 1149–1160. ▶ https://doi.org/10.3758/brm.41.4.1149.

# Determining Sample Size in Experimental Studies

© Springer Nature Singapore Pte Ltd. 2020
J. P. Verma and P. Verma, *Determining Sample Size and Power in Research Studies*,
https://doi.org/10.1007/978-981-15-5204-5_6

- **Learning Objectives**

After going through this chapter, the readers should be able to

— Describe the steps used in G*Power software for determining sample size and power in different applications.

— Know the procedure in computing sample size for a given power and effect size by using G*Power software.

— Explain the achieved power in hypothesis testing experiments if sample size is fixed.

— Determine the sample size or power in different types of $t$-tests for means and proportions using G*Power software.

— Decide the effect size in different hypothesis testing experiments.

— Estimate the sample size in correlation studies for a given effect.

— Know the procedure involved in estimating sample size and power in non-parametric tests.

**6**

## Introduction

We have seen formulas for determining sample size in one and two tailed tests. This gives an idea to the researchers as to how to determine sample size and power using precise mathematical formulas and how these formulas are derived. Chapter 6 and 7 have been developed for the applied researchers who wish to determine sample size or wish to know the power achieved in their experiments for answering their research questions using a statistical software. A common concern amongst researchers is that the sample is predetermined. In which case knowing and reporting the power of the test is very important. This chapter will enable researchers to calculate the power of the test employed in their studies. We have shown computation of power in the first few illustrations in this chapter. By using similar procedures, one can compute the power in other settings as well. In this chapter, we shall discuss the procedure of estimating sample size and power for different statistical tests by using G*Power software.

We have seen that in general five parameters are required for most of the statistical tests to determine the sample size. These are Type I error, power, minimum detectable difference, population variability and type of test (one/two-tailed test). We have examined the procedure of determining the sample size for desired power by fixing Type I error and effect size in hypothesis testing experiments. An experiment becomes meaningless if magnitude of the required effect is not decided in advance, i.e. how much minimum detectable difference (d) the researcher wishes to have if the testing results in rejection of the null hypothesis. This difference '$d$' is decided by the researcher so that in the event of rejecting the null hypothesis the claim may be meaningful for implementation. For detecting a smaller effect, a larger sample is required in comparison to that of larger effect in the experiment provided other conditions are same. After deciding the minimum detectable difference in the experiment, the next step is the estimation of population variance of the experimental variable. The population variance is usually unknown and is estimated by similar studies conducted earlier. Knowing variance is important because large variance requires a large sample for the same effect in comparison to the small variance provided other conditions are same.

In case the researcher is not able to decide as to how much minimum detectable difference should be taken or the estimate of population variability is not correctly known, then one can use Cohen's guidelines on effect size. Effect size is the ratio obtained by dividing the minimum detectable difference by the standard deviation. Different yardsticks for the effect size have been suggested by Cohen in different applications. For instance, in general, the effect size of 0.2 can be considered as low and 0.8 as high in hypothesis testing experiments, whereas in animal experiments range of low to high effect size may vary from 0.5 to 1.5. One may like to take the medium effect size for deciding the sample size.

The procedure discussed in this chapter for deciding sample size and power in different statistical tests may be helpful for checking the quality of thesis or validity of research findings by means of achieved power assuming effect size to be medium. If power in any study is less than 80%, one should cautiously read the findings. Similarly, if the power is 50% or less, then the results of the study may be considered no more than merely a guess.

## One Sample Tests

In this section we shall discuss the procedure of deciding sample size and computing power in studies concerning mean and proportion in one sample.

**One-sample $t$-Test for Mean**: Determining sample size and power.

- **Illustration 6.1**

The mean calcium concentration in a random sample of housewives in the age category 40–50 years is 2.5 mmol/l. What should be the sample size required for detecting the difference between the means of sample and population with power 0.8 at significance level 0.05 using one-tailed test, given that the mean and standard deviation of calcium concentration are 2.3 mmol/l and 0.5 mmol/l, respectively, in the population?

- **Solution**

*Determining sample size*

We need to estimate the sample size that is required to test whether the mean calcium concentration ($=2.5$ mmol/l) in a random sample of housewives is significantly different from the population mean of 2.3 mmol/l. The following information is given in the study:

$d$ = minimum detectable difference $= 0.2$ mmol/l
$\alpha = 0.05$
Power $= 0.8$
$\sigma = 0.5$ mmol/l
Type of test $=$ One tail

In order to determine the sample size, we shall first start the G*Power software after installation as described in ▶ Chap. 5. By following the below-mentioned steps as shown in ◻ Fig. 6.1, the sample size can be calculated.

Step 1: Select "$t$ tests" for two-sample test.
Step 2: Select "Means: Difference from constant".
Step 3: Select "A priori" to compute the sample size.

**6**

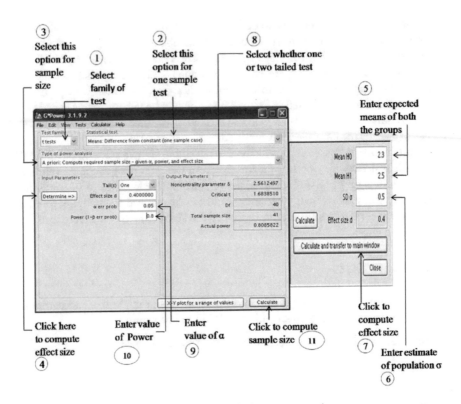

□ **Fig. 6.1**   Steps in G-power for determining sample size in one-sample test

Step 4: Click on ***Determine*** to compute the effect size in the adjacent window which pops up automatically.

Step 5: Enter the value of population and sample means in the location marked with Mean $H_0$ and Mean $H_1$, respectively.

Step 6: Enter population standard deviation.

Step 7: Click on ***Calculate and transfer to main window*** to compute the effect size and transfer its value in the main window.

Step 8: Choose one-tail test.

Step 9: Enter the value of $\alpha$ as 0.05.

Step 10: Enter the value of power as 0.8.

Step 11: Click on ***Calculate***.

***Remark*** In case population mean and variance are unknown or cannot be estimated from the previous studies, then one can directly use the effect size as per Cohen's guidelines (Low: 0.2, Medium: 0.5 and High: 0.8) instead of determining it in Step 4. In biological experiments where animals are used, the guidelines for choosing the effect size are different (Low: 0.5, Medium: 1.0 and High: 1.5). As a thumb rule, one may use medium effect size for determining sample size.

After pressing ***Calculate***, required sample size calculated in this study is 41 with achieved power of 80.86%.

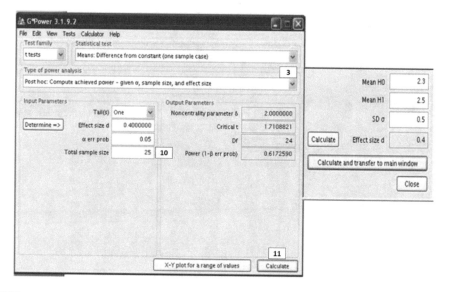

**□ Fig. 6.2**   Steps in G-power for determining power in one-sample test

**Inference**: A sample size of 41 is required for correctly rejecting the null hypothesis of no difference between the sample and population mean by 0.2 mmol/l of calcium concentration with 80.86% power at significance level 0.05 in one-tailed test.

*Determining power if the sample size is fixed*

In the above illustration if more than 25 subjects cannot be taken, let us see what would be the power in conducting the experiment.

In step 3, select option "Post hoc: Compute achieved power—given $\alpha$, sample size, and effect size" for determining the power. In step 10, enter the sample size as 25 and press **Calculate** in step 11 to determine the power in the window. All other steps would be same. The power for $n = 25$ when other parameters are as mentioned in the illustration shall be around 61.73% in the experiment as shown in □ Fig. 6.2.

**One-sample *t*-Test for Proportion**: Determining sample size.

- **Illustration 6.2**

A product manager in a biomedical company needs to detect whether the process deteriorates by more than 10% of defective pieces. He continues the process as long as it gives less than 10% defectives. He is interested in a one-sided test with significance level of 5%. What would be the sample size to detect the 10% further deterioration in comparison to the accepted 10% defectives with power 0.9?

- **Solution**

Here the following information is given:
  Proportion of defective pieces in population $(p_1) = 0.10$
  Proportion of defective pieces in sample $(p_2) = 0.20$

**6**

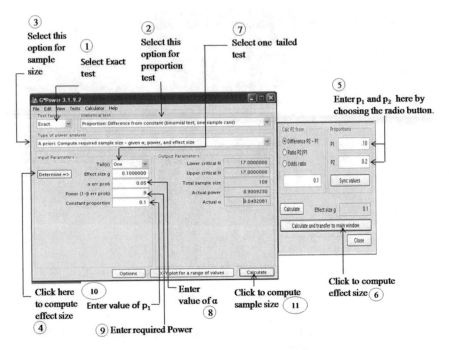

■ **Fig. 6.3**    Steps in G-power for determining sample size in one-sample proportion test

$\alpha = 0.05$
Power $= 0.9$
Type of test $=$ One tail

We need to determine the sample size n so as to detect the difference of 0.10 in proportion of defective pieces with 0.9 power. Let us follow the below-mentioned steps to determine the sample size using G*Power.

Step 1: Select "Exact" test for proportion.

Step 2: Select "Proportion: Difference from constant (binomial test, one sample case)".

Step 3: Select "A priori" to compute the sample size.

Step 4: Click on **Determine** to compute the effect size in the adjacent window which pops up automatically.

Step 5: Enter the values of $p_1$ and $p_2$ in the respective windows by choosing the radio button "Difference $p_2 - p_1$".

Step 6: Click on **Calculate and transfer to main window** to compute the effect size and transfer its value in the main window.

Step 7: Choose one-tail test.

Step 8: Enter the value of $\alpha$ as 0.05.

Step 9: Enter the value of power as 0.9 (■ Fig. 6.3).

Step 10: Enter the value of $p_1$ (proportion in the population).

Step 11: Click on **Calculate** for sample size.

After pressing **Calculate**, required sample size and power actually achieved are calculated. In this case, estimated sample size is 109.

**Inference**: A sample of size 109 is required for detecting the 0.1 difference in proportion of defectives between the sample and population proportions with 90% power tested at the significance level of 0.05 in one-tailed test.

## Two Sample Tests

In many situations it is desired to compare the means or proportions of some characteristics in two groups. If data comes from the normal population then two sample t-test or paired t-test are used to compare means of the groups depending upon whether the groups are independent or related respectively. When normality violates, group mean comparisons are done using nonparametric tests such as Mann-Whitney U test or Wilcoxon signed-rank test. We shall discuss the computation of sample size and power in using all these statistical tests.

**Two-sample _t_-Test for Mean**: Determining sample size and power.

- Illustration 6.3

What sample size is needed in correctly rejecting the null hypothesis of no difference in the noise level in the Lobby and Library in a college with power 0.8? The expected mean noise level for the Lobby is 62.4 dBA and for the Library is 58.1 dBA. The expected standard deviation (SD) in each group is 7.8 dBA. One-tailed test is proposed to be applied at the significance level of 0.05.

Further, if in the lobby, 35 sample data could be collected, what sample size would be required in the library to have the same power in the experiment if the other conditions are constant?

- Solution

_Determining sample size_

Here, the expected standard deviation of noise level in the population is given as 7.8 dBA which is same in both the locations as this is one of the assumptions of using t-test as well. This knowledge of the standard deviation might be obtained from the review of literature or it can be obtained by conducting a pilot study. The following information is provided in the study:

$\alpha = 0.05$

Power $= 0.8$

Mean Group 1 $= 62.4$

Mean Group 2 $= 58.1$

$S = 7.8$ [sd of both the groups are same]

Type of test $=$ One tail

Follow the below-mentioned steps to estimate the sample size as shown in ◻ Fig. 6.4 in a sequential manner.

Step 1: Select "_t_ tests" for two-sample test.

Step 2: Select option "Means: Difference between two independent means".

6

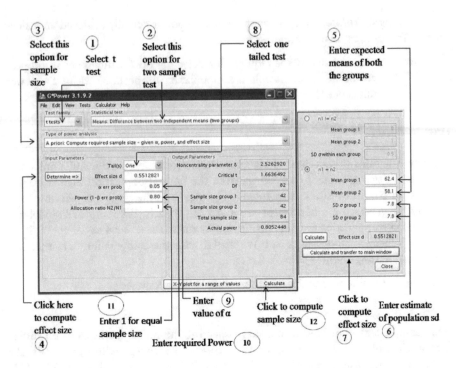

**◻ Fig. 6.4**    Steps in G-power for determining sample size in comparing two independent sample means

Step 3: Select "A priori" to compute the sample size.
Step 4: Click on **Determine** to compute the effect size in the adjacent window, which
pops up automatically.

**Remark**
1. There are two methods of determining the effect size:
   *Method 1* (Unbalanced Design): Here, the sample size in both the groups is
   expected to be different. In this case, select the option "$n1 \ne n2$".
   *Method 2* (Balanced Design): Here, the samples in both the groups are expected
   to be the same, select the option "$n1 = n2$".
   After choosing one of the above options, enter the means of both the groups.
   Enter standard deviation. If you have preliminary data, you can compute pooled
   standard deviation and then use that value.
2. We have chosen Method 2 as the sample size is expected to be the same here.
   Step 5: Enter the means of both the groups as 62.4 and 58.1.
   Step 6: Enter the estimate of population standard deviation as 7.8.
   [**Remark**: If mean and standard deviation for the data obtained in both
   the groups are not available from the earlier studies, then the researcher
   may choose the required effect size as per Cohen's guidelines (Low: 0.2,
   Medium: 0.5 and High: 0.8). In biological experiments where animals
   are used, the guidelines for choosing the effect size are different (Low:
   0.5, Medium: 1.0 and High: 1.5)].

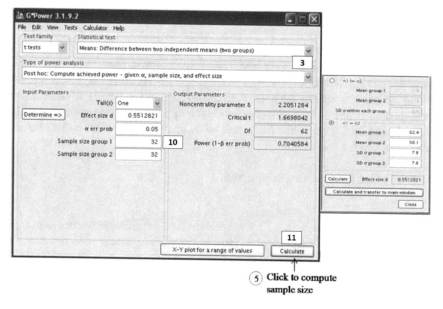

5 Click to compute
sample size

☐ **Fig. 6.5**  Steps in G-power for determining power in two-sample $t$-test

Step 7: Click on **Calculate and transfer to main window** to compute the effect
size in the study and transfer its value in the main window.

Step 8: Choose one-tail test.

Step 9: Enter the value of $\alpha$ as 0.05.

Step 10: Enter power as 0.8.

Step 11: Enter Allocation ratio as 1 (If expected sample sizes in both the groups
are same).

Step 12: Click on **Calculate**.

After pressing **Calculate**, the sample size and power actually achieved are calculated in the adjacent column. Here, the sample size ($n$) required for each group is 42
for each sample.

If in the lobby only 35($n_1$) sample data could be collected, then we need to compute the sample size ($n_2$) required in the library to have the same power provided
other conditions are same.

Here, $n = 42$ and $n_1 = 35$, $n_2 = ?$

$$n_2 = nn_1/(2n_1 - n) = (42 \times 35)/(2 \times 35 - 42) = 52.5$$

Thus, if $n_1 = 35$ then $n_2$ should be 53 to have the same power provided other conditions are same.

**Inference** The sample size in each group should be 42 in order to detect the difference
in noise level of 4.3 dBA in lobby and library with 80.52% power tested at 5% level
in one-tailed test. If in the lobby only 35 data could be collected, then in the Library
53 data should be collected to have the same chance of correctly rejecting the null
hypothesis.

*Determining power if the sample size is fixed*
In illustration 6.3, if more than 32 subjects cannot be taken in each sample let us see what would be the power in the experiment (❑ Fig. 6.5).

In step 3, select option "Post hoc: Compute achieved power—given $\alpha$, sample size, and effect size" for determining the power. In step 10, enter the sample size as 32 because sample size is same here and then press **Calculate** in step 11 to determine the power in the window. All other steps remain the same. The power of the test with 32 sample data in each group shall be around 0.704 if other parameters remain the same as mentioned in the illustration.

**Paired *t*-Test**: Determining Sample Size.

- **Illustration 6.4**

An exercise scientist prepares a 12-week weight loss programme for the housewives who are in 30–40 years' age category. How many participants should constitute the sample to detect 80% times a reduction of 3 kg in weight with significance level 0.05? The expected average weight of the participants before joining the programme is 90 kg, and expected standard deviation is 5 kg. The expected average weight of the same participants after the 12-week programme is 87 (expected standard deviation $= 5.5$ kg).

- **Solution**

In this illustration, the following parameters are provided:
$\alpha = 0.05$
Power $= 0.8$
Mean weight in pre-testing $= 90$ kg
SD of weight in pre-testing $= 5$ kg
Mean weight in post-testing $= 87$ kg
SD of weight in post-testing $= 5.5$ kg
Type of test $=$ One tail
Follow the below-mentioned steps to estimate the sample size as shown in ❑ Fig. 6.6 in a sequential manner.
Step 1: Select "*t* tests" for paired sample.
Step 2: Select option "Means: Difference between two dependent means".
Step 3: Select "A priori" to compute the sample size.
Step 4: Click on **Determine** to compute the effect size in the adjacent window which pops up automatically.
Step 5: Enter mean of both the groups as 90 and 87.
Step 6: Enter the estimate of population standard deviations in two groups as 5 and 5.5.
   [**Remark** If mean and standard deviation for the data obtained in the pre- and post-testing are not available from the earlier studies, there are two options for the researchers. Firstly, he may decide how much mean difference he is expecting in pre- and post-testing results and estimate the standard deviation of the difference data in pre- and post-testing by the pilot study, alternatively he should choose the required effect size as per Cohen's guidelines (Low: 0.2, Medium: 0.5 and High: 0.8) and enter directly into the window for calculating sample size.]

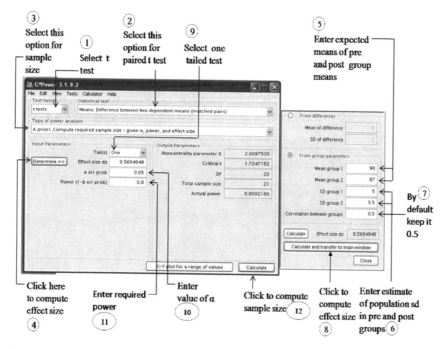

**☐ Fig. 6.6** Steps in G-power for determining sample size in paired *t*-test

Step 7: Enter moderate correlation between two groups. By default it may be taken as 0.5 if the information about the correlation between the two populations is unknown.

Step 8: Click on **Calculate and transfer to main window** to compute the effect size and transfer its value in the main window.

Step 9: Choose one-tail test.

Step 10: Enter the value of $\alpha$ as 0.05.

Step 11: Enter power as 0.8.

Step 12: Click on **Calculate**.

After pressing **Calculate**, the sample size and power actually achieved are calculated in the windows. Here, the required sample size ($n$) is 21.

**Inference** With sample size 21, there are 80.1% chances to detect the difference of 3 kg reduction in weights of housewives during 12-week weight loss programme in a one-tailed test with significance level 0.05.

**Mann–Whitney Test:** Determining Sample Size.

The procedure discussed in Illustration 6.5 is meant for metric data that do not satisfy parametric assumptions and is not meant for the ordinal data. In fact, procedure similar to that of parametric '*t*' test is used in this situation for sample size and power determination. The only difference is that in this case the required sample size will usually be higher.

- **Illustration 6.5**

In order to test whether study hours of women are higher than that of men in a college, a researcher finds that '*t*'-test cannot be used as assumption of equal variance in two groups is violated. Hence, he decides to use the Mann–Whitney test for comparing these two groups. What sample size should be taken to detect the difference of 0.25 h between the group means with power 0.9 and $\alpha = 0.05$. The expected mean and standard deviation for women and men in the experiment are shown in the following table:

*Estimated means and standard deviation*

|      | Women   | Men     |
| ---- | ------- | ------- |
| Mean | 6 h     | 5.75 h  |
| SD   | 0.5 h   | 0.5 h   |

- **Solution**

We need to determine the required sample size for detecting a difference of 0.25 h with 0.9 power in this illustration. The following information is given:

$\alpha = 0.05$
Power $= 0.9$
Mean of the first group $= 5.75$
Mean of the second group $= 6$
$S = 0.5$ [sd of both the groups are same]
Type of test $=$ One tail

By following the below-mentioned steps as shown in ◻ Fig. 6.7, the sample size can be calculated.

Step 1: Select "*t* tests".
Step 2: Select option "Means: Wilcoxon–Mann–Whitney test".
Step 3: Select "A priori" to compute the sample size.
Step 4: Click on **Determine** to compute the effect size in the adjacent window which pops up automatically. Choose "$n_1 = n_2$" option in pop window.
Step 5: Enter mean of both the groups as 5.75 and 6.
Step 6: Enter the estimate of population standard deviation as 0.5.

> [**Remark** If mean and standard deviation for the data obtained in both the groups are not available from the earlier studies, then the researcher may choose the required effect size as per Cohen's guidelines (Low: 0.2, Medium: 0.5 and High: 0.8).]

Step 7: Click on **Calculate and transfer to main window** to compute the effect size in the study and transfer its value in the main window.
Step 8: Choose one-tail test.
Step 9: Choose Parent distribution as Laplace.

> [**Remark** Choose Laplace if the distribution is leptokurtic and Logistic if platykurtic. Usually when the distribution is highly skewed then Mann–Whitney test is preferred. Hence, by default one can choose Laplace.]

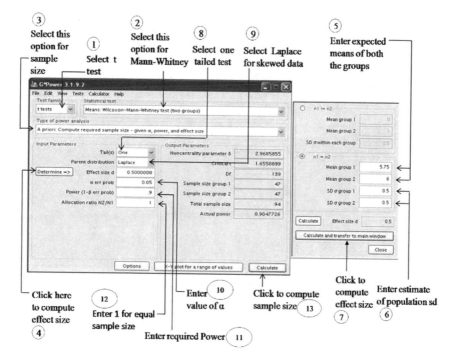

**Fig. 6.7** Steps in G-power for determining sample size in Mann–Whitney test

Step 10: Enter the value of $\alpha$ as 0.05.
Step 11: Enter power as 0.9.
Step 12: Enter Allocation ratio as 1 (If expected sample sizes in both the groups are same).
Step 13: Click on **Calculate**.

After pressing **Calculate**, the sample size and the power actually achieved are calculated in the adjacent blocks. Here, the sample size ($n$) required for each group is 47 for each sample.

**Inference** The sample size in each group should be 47 to detect the difference of 0.25 in study hour of women and men with 90.5% power and significance level 0.05 in one-tailed test.

**Wilcoxon Signed-Rank Test**: Determining Sample Size.

The procedure discussed in Illustration 6.6 is meant for metric data that do not satisfy parametric assumptions and is not meant for the ordinal data. In fact procedure similar to that of parametric $t$-test is used in this situation for sample size and power determination. The only difference is that in this case the required sample size is usually higher.

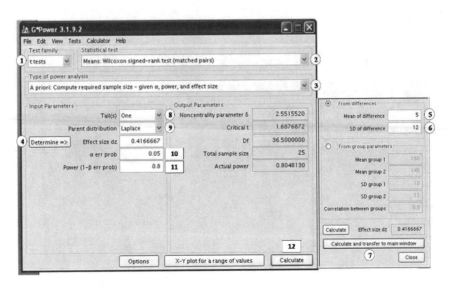

**□ Fig. 6.8** Steps in G-power for determining sample size in Wilcoxon signed-rank test for paired groups

- **Illustration 6.6**

In studying the effect of a drug on blood sugar, it was found that the paired '*t*'-test cannot be used as assumption of normality for the difference of scores is violated. Hence, it was decided to use the Wilcoxon signed-rank test for paired data. What sample size should be taken to detect a difference of 5 counts between group means with 80% power at significance level 0.01? Earlier studies suggest that the *SD of the difference ($d_i$) is 12 units.*

- **Solution**

We need to determine the required sample size when power is 0.8. The following information is provided:

   Mean of difference (minimum detectable difference) = 5
   SD of difference = 12
   Type of test = One tail
   $\alpha = 0.01$
   Power = 0.8

By following the below-mentioned steps as shown in □ Fig. 6.8, the sample size can be calculated.

Step 1: Select "t tests".
Step 2: Select option "Means: Wilcoxon signed-rank test (matched pairs)".
Step 3: Select "A priori" to compute the sample size.
Step 4: Click on **Determine** to compute the effect size in the adjacent window which pops up automatically.
Step 5: Enter mean of difference as 5.

Step 6: Enter the SD of difference as 12.

> [*Remark* If the researcher cannot decide as to how much difference is expected in the pre- and post-testing or standard deviation of the difference data in pre- and post-testing cannot be estimated, he may choose the required effect size as per Cohen's guidelines (Low: 0.2, Medium: 0.5 and High: 0.8).]

Step 7: Click on **Calculate and transfer to main window** to compute the effect size and transfer its value in the main window.

Step 8: Choose one-tail test.

Step 9: Choose parent distribution as Laplace.

> [*Remark* Choose Laplace if the distribution is leptokurtic and Logistic if platykurtic. Usually, when the distribution is highly skewed then signed-rank test is preferred. Hence, by default one can choose Laplace.]

Step 10: Enter the value of $\alpha$ as 0.05.

Step 11: Enter power as 0.8.

Step 12: Click on **Calculate.**

After pressing **Calculate**, the sample size and achieved power are calculated in the window. Here, the required sample size $(n)$ is 25.

**Inference** There are 80.48% chances in correctly rejecting the null hypothesis that the drug is not effective with sample size 25 and significance level 0.01 in one-tailed test.

***T*-test for Comparing Two Proportions:** Determination Sample Size.

- **Illustration 6.7**

Earlier studies show that the expected proportion of child vaccination in northern cities is 0.9 and in the southern cities it is 0.85. How many cases are needed to correctly reject the null hypothesis of no difference between the proportions of child vaccination in northern and southern cities 80% times at the significance level of 0.05? Due to difference in population size, it is decided to have the proportion of samples $n_2/n_1$ to be 1.25 where $n_1$ is the sample size in northern cities and $n_2$ in southern cities.

- **Solution**

In this illustration, the following parameters are given:

$\alpha = 0.05$

Power $= 0.8$

$P_1$(proportion in northern cities) $= 0.9$

$P_2$(proportion in southern cities) $= 0.85$

Type of test $=$ Two tail

By following the below-mentioned steps as shown in ◻ Fig. 6.9, the sample size can be calculated.

Step 1: Select "Exact" test for comparing proportions of two groups.

Step 2: Select option "Proportions Inequality, two Independent groups".

Step 3: Select "A priori" to compute the sample size.

**6**

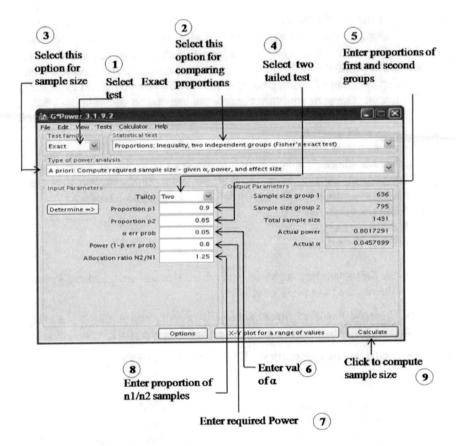

□ **Fig. 6.9** Steps in G-power for determining sample size in comparing proportions in two groups

Step 4: Choose two-tail test.
Step 5: Enter proportions of two groups as $p_1 = 0.9$ and $p_2 = 0.85$.
Step 6: Enter the value of $\alpha$ as 0.05.
Step 7: Enter power as 0.8.
Step 8: Enter the proportion of sample 1.25. This indicates that due to different population sizes it is required to have a sample in southern cities 1.25 times that of northern cities.
Step 9: Click on **Calculate**.

After pressing **Calculate**, the sample sizes are calculated for both the groups along with the achieved power. Here, required sample size in northern districts ($n_1$) is 636 and that of in southern districts ($n_2$) is 795.

**Inference** Samples of size 636 in northern districts and 795 in southern districts are needed to detect a difference of 0.05 in proportions in the districts of two regions with 80% power and significance level 0.05 in a two-tailed test.

# Testing Significance of Relationship

In this section we shall discuss the procedure for determining the sample size in testing the significance of correlation coefficient, difference in two correlations and biserial correlation. Further, estimating sample size in testing goodness of fit has also been discussed by means of an illustration.

**Testing the Significance of Correlation Coefficient**: Determining Sample Size.

- **Illustration 6.8**

In testing the significance of correlation coefficient between resting respiratory rate and fat%, what sample size should be taken to test whether the observed sample correlation differs from 0 and having at least 45% coefficient of determination with 0.9 power and significance level 0.05 in carrying out a two-tailed test?

- **Solution**

In this illustration, we wish to determine the sample size for testing the significance of observed correlation coefficient between resting respiratory rate and fat% in detecting 45% coefficient of determination.

Coefficient of determination is the square of correlation coefficient which explains the percentage variability in dependent variable explained by the independent variable. Here, the coefficient of determination is 45% and we would like to determine the significance of observed correlation coefficient, i.e. whether the observed correlation coefficient is significantly different from 0 and explains at least 45% variability in resting respiratory rate by the fat%.

Here, we are given the following information:

Coefficient of determination $= 0.45$

Type of test $=$ Two tail

$\alpha = 0.05$

Power $= 0.9$

Correlation coefficient under $H_0 = 0$

Following the below-mentioned steps will determine the required sample size. The procedure has been shown graphically in ◻ Fig. 6.10.

Step 1: Select "Exact" tests for correlation coefficient.

Step 2: Select option "Correlation: Bivariate normal model".

Step 3: Select "A priori" to compute the sample size.

Step 4: Click on **Determine** to compute the correlation you expect in sample for being significant, in the adjacent window which pops up automatically.

Step 5: Enter the Coefficient of determination you wish to have. Here it is 0.45.

> [**Remark** The coefficient of determination is the effect which the researcher is interested to test. In case if the researcher cannot decide as to how much $R$-square he should expect then he may choose the required effect size for Pearson's correlation ($r$) as per Cohen's guidelines (Low: 0.1, Medium: 0.3 and High: 0.5).]

Step 6: Click on **Calculate and transfer to main window** to compute effect size and transfer its value in the main window.

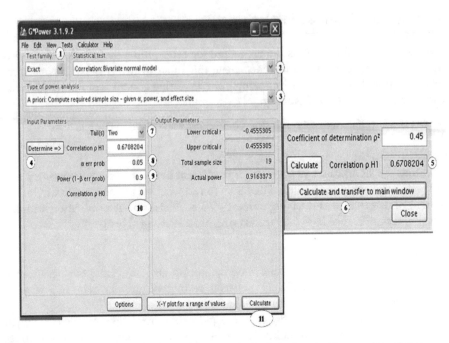

□ **Fig. 6.10**  Steps in G-power for determining sample size in testing significance of correlation coefficient

Step 7: Enter type of test as two tail.
Step 8: Enter $\alpha$ as 0.05.
Step 9: Enter power as 0.9.
Step 10: Enter correlation under $H_0$ as 0.
Step 11: Click on **Calculate**.

After pressing **Calculate**, the sample size calculated for testing the significance of correlation coefficient is 19 for achieving power 0.9163.

**Inference** Thus, a sample of 19 paired observations is required to detect the effect of 0.45 $R^2$ with 91.63% power and significance level 0.05 in a two-tailed test. In other words, a sample of 19 paired observations is required to reject the null hypothesis that the correlation between resting pulse rate and fat% is not significant, correctly 91.63% time at significance level 0.05 in a two-tailed test.

**Testing whether Correlation Differs from a Known Value**: Determining Sample Size.

- **Illustration 6.9**
Correlation between height and weight is known to be 0.55 among children in the age category of 8–12 years. What should be the sample size required in correctly rejecting the null hypothesis that the sample correlation of 0.7 differs from 0.55, correctly 80% time at significance level 0.05 in a two-tailed test?

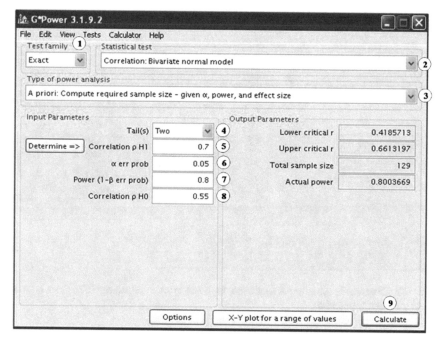

**◻ Fig. 6.11** Steps in G-power for determining sample size in testing significance of correlation coefficient with a reference value

■ **Solution**

Here, we have the following information:

Type of test = Two tail

Correlation under $H_1 = 0.7$

$\alpha = 0.05$

Power = 0.8

Correlation under $H_0 = 0.55$

Following the below-mentioned steps will determine the required sample size. The procedure has been shown graphically in ◻ Fig. 6.11.

Step 1: Select "Exact" tests for correlation coefficient.

Step 2: Select option "Correlation: Bivariate normal model".

Step 3: Select "A priori" to compute the sample size.

Step 4: Enter type of test as two tailed.

Step 5: Enter correlation under $H_1$ as 0.7.

Step 6: Enter $\alpha$ as 0.05.

Step 7: Enter power as 0.8.

Step 8: Enter correlation under $H_0$ as 0.55.

Step 9: Click on **Calculate**.

After pressing **Calculate**, the sample size calculated for testing the significance of the difference in correlation coefficients and detecting a difference of 0.15 is 129 with achieved power 0.8.

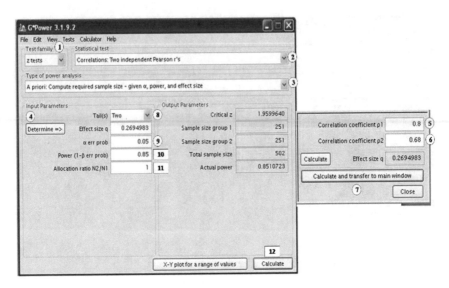

**◻ Fig. 6.12** Steps in G-power for determining sample size in testing significant difference between two independent correlations

**Inference** A sample of size 129 is required to reject the null hypothesis of no difference in correlation coefficients, correctly 80% time at significance level 0.05 in a two-tailed test.

**Testing Significant Difference between two Correlations**: Determining sample size.

- **Illustration 6.10**

In a correlational study between IQ and Math's score, how many sample data would be required in correctly rejecting the null hypothesis of equal correlations in men and women if the following information is given?
  Estimated correlation in men $(\rho_1) = 0.8$
  Estimated correlation in women $(\rho_2) = 0.68$
  Type of test $=$ two tail
  Significance level $(\alpha) = 0.05$
  Power $(1 - \beta) = 0.85$
  Sample sizes in both the groups are same, i.e. $n_1/n_2 = 1$.

- **Solution**

With the above information, let us calculate the required sample size. Below-mentioned steps in G*Power software will determine the required sample size. The procedure has been shown graphically in ◻ Fig. 6.12.
Step 1: Select "Z tests".
Step 2: Select option "Correlations: Two independent Pearson r's".
Step 3: Select "A priori" to compute the sample size.
Step 4: Click on **Determine** to compute the effect size in the adjacent window which pops up automatically.
Step 5: Enter correlation coefficient for the men $(\rho_1) = 0.8$.

Step 6: Enter correlation coefficient for the women $(\rho_2) = 0.68$.

Step 7: Click on **Calculate and transfer to main window** to compute effect size and transfer its value in the main window.

Step 8: Enter type of test as two tail.

Step 9: Enter $\alpha$ as 0.05.

Step 10: Enter power $(1 - \beta)$ as 0.85.

Step 11: Enter allocation ratio $n_1/n_2 = 1$.

[This is because we intend to take the same size sample.]

Step 12: Click on **Calculate**.

After pressing **Calculate**, the sample size calculated for testing the significance difference between the two correlations is 502 for achieving 0.8511 power.

**Inference** There are 85.11% chances of correctly rejecting the null hypothesis of no difference between the men and women correlations (in IQ and math's score) with 251 sample data in each gender with significance level 0.05 in a two-tailed test.

**Testing Significance of Bi-serial Correlation**: Determining sample size

We shall now examine the procedure in determining the required sample size in testing the significance of bi-serial correlation with the help of the following illustration.

**■■ Illustration 6.11**

A researcher is interested to test whether haemoglobin content is gender specific in 40–50 years' age category. Because one of the variables (Gender) is dichotomous and the other variable (haemoglobin) is numeric, the use of bi-serial correlation is decided; correlation is expected to be at least 0.6. What sample size should be selected to test whether the observed correlation is significantly greater than 0 with power 0.9 while using one-tailed test at the significance level 0.05?

**■ Solution**

In this illustration, we need to determine the sample size required for testing the hypothesis that the bi-serial correlation is significant with 0.9 power and 0.05 significance level. The following information is given:

Level of significance $= 0.05$

Required power $= 0.90$

Type of test $=$ One tail

Expected coefficient of determination $(\rho^2) = 0.36$

By performing the following steps in G*Power software, the required sample size can be determined. The procedure has been shown sequentially in ◘ Fig. 6.13.

Step 1: Select "t tests".

Step 2: Select option "Correlation: Point biserial model".

Step 3: Select "A priori" to compute the sample size.

Step 4: Click on **Determine** to compute the effect size in the adjacent window which pops up automatically.

Step 5: Enter the Coefficient of determination $(\rho^2)$ as 0.36.

6

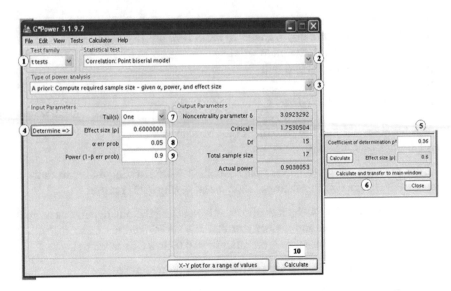

□ **Fig. 6.13**   Steps in G-power for determining sample size in testing the significance of Bi-serial correlation

**Remark**

1. Coefficient of determination is the amount of variability in dependent variable explained by the independent variable. Thus, in this illustration, we wish that around 36% variability in haemoglobin should be explained by the gender in testing.
2. If it is difficult to estimate the coefficient of determination ($\rho^2$) from earlier studies, one can follow the below-mentioned guidelines.
   Guidelines for effect size in point bi-serial correlation

| Effect size ($w$) | Magnitude |
|---|---|
| 0.1 | Low |
| 0.3 | Medium |
| 0.5 | High |

3. If you wish to detect a smaller effect, a larger sample is required in comparison to the larger effect for the same power.

   Step 6: Click on **Calculate and transfer to main window** to compute effect size and transfer its value in the main window.
   Step 7: Enter type of test as one tail.
   Step 8: Enter $\alpha$ as 0.05.
   Step 9: Enter power as 0.9.
   Step 10: Click on **Calculate**.

After pressing **Calculate**, the sample size calculated for testing the significance of bi-serial correlation is 17 for achieving 0.9038 power.

**Inference** There are 90.38% chances of correctly rejecting the null hypothesis that the bi-serial correlation is equal to zero with 17 samples at 5% level in a one-tailed test.

**Goodness-of-Fit Testing with Chi-Square**: Determining Sample Size.

■ **Illustration 6.12**

An epidemiologist is interested to know whether incidence of four diseases (Malaria, Dengue, Chikungunya and Bird flu) in a state is equal to the national proportions (0.4, 0.2, 0.3 and 0.1, respectively). The proportions of incidences of the four diseases Malaria, Dengue, Chikungunya and Bird flu in the state are 0.50, 0.15, 0.15 and 0.20, respectively. What should be the sample size to test the goodness of fit with 0.9 power and 0.05 significance level?

■ **Solution**

Here, the following information is given:

$\alpha = 0.05$

Power $= 0.90$

Degrees of freedom $= 3$ (Total number of groups $- 1 = 4 - 1$)

*Observed and expected proportions*

|  | Expected proportion $p(H_0)$ | Observed proportion $p(H_1)$ |
|---|---|---|
| Malaria | 0.4 | 0.5 |
| Dengue | 0.2 | 0.15 |
| Chikungunya | 0.3 | 0.15 |
| Bird flu | 0.1 | 0.20 |

By performing the following steps, the required sample size can be determined. The procedure has been shown sequentially in ◘ Fig. 6.14.

Step 1: Select "$\chi^2$ tests" for goodness of fit.

Step 2: Select option "Goodness-of-fit tests: Contingency tables".

Step 3: Select "A priori" to compute the sample size.

Step 4: Click on **Determine** to compute the effect size in the adjacent window which pops up automatically.

Step 5: Enter proportions of different diseases in population, i.e. $p(H_0)$ as 0.4, 0.2, 0.3 and 0.1. (Since the diseases in the population are in these proportions.)

Step 6: Enter proportions of different diseases in sample, i.e. $p(H_1)$ as 0.50, 0.15, 0.15 and 0.20. (Since the diseases in the sample are in these proportions.)

Step 7: Click on **Calculate and transfer to main window** to compute the effect size and transfer its value in the main window.

  **Remark** The effect size w is calculated by the following formula: $w = \sum \frac{(p_0 - p_1)^2}{p_0}$

Step 8: Enter value of $\alpha$ as 0.05.

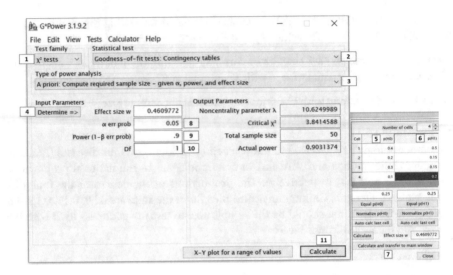

**◘ Fig. 6.14** Steps in G*Power for determining sample size in testing the goodness of fit

Step 9: Enter power as 0.9.

Step 10: Enter degrees of freedom as 1.

> **Remark** Degrees of freedom is obtained by the formula $m$-$r$-1, where $m$ is the number of categories, $r$ is 2 as two degrees of freedom is lost in estimating mean and variance of the distribution of probabilities. If probabilities under null hypothesis are same for each category, then the degrees of freedom would be obtained by $m - 1$. Thus in this case degrees of freedom is $1(m - r - 1 = 4 - 2 - 1)$.

Step 11: Click on **Calculate**.

After pressing **Calculate**, the sample size is determined. Here, required sample size is 50 for achieving power 0.9015.

**Inference** There are 90.15% chances of correctly rejecting the null hypothesis of no difference between expected and observed proportions with 50 subjects.

## Summary

Sample size can be calculated in different applications like all types of $t$-tests for mean and proportion; non-parametric tests like Mann–Whitney, Wilcoxon signed-rank; testing significance of correlations, comparing correlations and bi-serial correlation; and Goodness-of-fit tests. Similarly, for a given sample size, achieved power can be computed in these applications. In using the $t$-test for single group mean if one cannot decide as to how much minimum detectable difference should be there or population variability is difficult to estimate then Cohen's guidelines for effect size can be used. As per his guidelines, 0.2 is considered as low effect size, 0.5 as medium and 0.8 as large. In biological experiments, the guidelines for choosing the effect size are different

(Low: 0.5, Medium: 1.0 and High: 1.5). The researcher may use medium effect size to determine the required sample size. The guidelines for deciding effects size depend upon the statistical test used in the study which have been discussed in different applications in the chapter.

■ Exercises

1. What is the importance of reporting effect size? Can it be greater than 1?

2. In testing whether IQ of the students is greater than 85, if the null hypothesis is rejected at 1% level without reporting effect size and power, what conclusion can be drawn?

3. To compare the effectiveness of a drug on men and women, the G*Power software suggests that 128 subjects are required (64 men and 64 women) in a two-tailed test for detecting the effect size of 0.5 with power 0.8 at the significance level 0.05. But while planning the experiment only 50 women subjects were available, what should be the number of men subjects you should have to ensure the same power in testing?

4. A researcher wishes to verify whether mean IQ of the students joining a university has risen to 85 from 81 reported earlier, after advertising its programme at the national level. If standard deviation is 9, what sample size would be required to detect the increase in mean IQ by 4 score with power 0.8 at 5% level given that the IQ is normally distributed?
   If the researcher could take only 25 samples, what would be the power in correctly rejecting the null hypothesis if all other conditions remain the same?

5. An analyst in a bulb manufacturing company needs to detect whether the process deteriorates by more than 2% of defective pieces. He continues the process as long as it gives less than 2% defectives. In a one-tailed test with 5% level, what would be the sample size to detect the 2% further deterioration in comparison to the accepted 2% defectives with power 0.8?
   If the researcher has only 40 samples, what would be the power in correctly rejecting the null hypothesis if all other parameters are same?

6. It is desired to compare the performance on vocabulary test of the subjects in 20 year's and 50 year's age categories. The scores on vocabulary test are assumed to be normally distributed. The scores have equal standard deviation in both the groups which is equal to 8. What should be the sample size to detect the difference in the means of two groups by 5 scores with 90% power at 5% level in a one-tailed test?
   If number of subjects in 20 year's group is only 35, what should be the number of subjects in 60 year's group to have the same power?

7. In an experiment, participants threw darts at a target first by using their preferred hand and then by using other hand. All subjects performed in both the conditions. The standard deviation of the difference between scores of the two groups is 9.5. What should be the sample size to detect the difference of 5 scores in the mean performance by both the hands with power 0.9 at significance level 0.05 in a one-tailed test?

8. A cricket analyst is interested to know whether mean batting average of Rattlers and Vikings differs significantly in league tournaments. The league batting

performance is known to have a standard deviation of 28 and violates normality; hence, the analyst decides to apply Mann–Whitney test to compare the batting averages. What should be the sample size of Rattlers and Vikings to detect the difference in their batting performance by 18 scores with 80% power at 1% level in a two-tailed test?

9. A medical researcher is interested to test the effectiveness of a new medicine in improving sleep hour among elder population. The sleep hour will be recorded before and after administering the medicine. Since paired $t$-test cannot be used as normality violates for the data on sleep, hence, he decides to use the Wilcoxon signed-rank test. What sample size should he take to detect the difference between the pre- and post-sleep hours by 0.5 h with 90% power at significance level 0.05 if one-tailed test is to be used. Earlier studies suggest that the SD of the difference ($d_i$) is 0.9 h?

10. It is believed that the proportion of men qualifying their drivers' test in the first attempt is 0.8 and that of women is 0.7. What should be the size of the sample to detect the difference between the two proportions with 80% power at significance level 0.05 in a two-tailed test? Due to difference in population size, it is decided to have the proportion of samples men to women to be 1.2.

11. In testing the correlation between body mass index (BMI) and weight, what should be the sample size in correctly rejecting the null hypothesis of no correlation with power 0.8 in detecting $R^2$ of 0.35 at significance level 0.01 in a two-tailed test?

12. It is known that correlation between the number of hours studied and number of correct answers in a mathematics test is 0.75. What sample size should be taken in correctly rejecting the null hypothesis that the sample correlation of 0.82 differs from 0.75 with power 0.9 at significance level 0.05 in a two-tailed test?

13. A researcher computes correlation between the scores of reasoning and memory retention in boys and girls in a school. He desires to test whether correlation values differ between boys and girls. What should be the size of the sample in correctly rejecting the null hypothesis of no difference between the correlations with 80% power in a two-tailed test if the following information is given?
  Estimated correlation in boys ($\rho_1$)=0.72
  Estimated correlation in girls ($\rho_2$)=0.60
  Significance level ($\alpha$)=0.05
  Sample sizes in both the groups are same, i.e. $n_1/n_2=1$

14. A researcher is interested to test whether minimum muscular fitness test depends upon the fat% in adult men. He plans to use bi-serial correlation because minimum muscular fitness test (pass vs. fail) is dichotomous and fat% is numeric. The estimated correlation is 0.4. How much sample size he should take in correctly rejecting the null hypothesis that the observed correlation is significantly greater than 0, with power 0.85 at 5% level in a two-tailed test?

15. It is known that the proportions of people having blood types AB, B, A and O in a district are known to be in the ratio 8: 12: 25: 45, respectively. A sample of subjects selected from the district shows frequencies of the blood types AB, B, A, O in the ratio 5: 15: 45: 85. What should be the sample size to test whether the sample proportions differ significantly from the population proportions with power 0.8 at significance level 0.01?

- **Answer**
  3. Number of men $= 89$
  4. $n = 33$, power $= 69.63\%$
  5. (i) $n = 35$
     (ii) power $= 87.14\%$
  6. (i) Each group should have 45 sample data
     (ii) If $n_1 = 35$ then $n_2$ should be 63
  7. $n = 33$
  8. $n_1 = 53$ and $n_2 = 53$
  9. $n = 20$
  10. $n_1(\text{men}) = 278$, $n_2(\text{women}) = 334$
  11. $n = 24$
  12. $n = 314$
  13. $n_1 = 272$, $n_2 = 272$
  14. $n = 50$
  15. $n = 201$

# Bibliography

Alotaibi, S. R. D., Roussinov, D. (2016). Using GPower software to determine the sample size from the pilot study. In *The 9th Saudi Students Conference, 2016-02-13*, University of Birmingham.

Bonett DG, Wright TA. (2000). Sample size requirements for estimating. Pearson, Kendall and Spearman correlations. *Psychometrika, 65*, 23–28.

Bujang, M. A., & Baharum, N. (2016). Sample size guideline for correlation analysis. *World Journal of Social Science Research. 3.* 37. ▶ https://doi.org/10.22158/wjssr.v3n1p37.

Bujang, M. A., & Baharum, N. (2016). Sample size guideline for correlation analysis. *World Journal of Social Science Research, 3*(1). ISSN 2375-9747 (Print) ISSN 2332-5534 (Online).

Collings, B. J., & Hamilton, M. A. (1998). Estimating the power of the two-sample Wilcoxon test for location shift. *Biometrics, 44*, 847–860.

Comulada, W. S., & Weiss, R. E. (2010). Sample size and power calculations for correlations between bivariate longitudinal data. *Statistics in Medicine, 29*(27), 2811–2824. ▶ https://doi.org/10.1002/sim.4064.

Dupont, W. D. (1988). Power calculations for matched case-control studies. *Biometrics, 44*, 1157–1168.

Faul, F., Erdfelder, E., Buchner, A., & Lang, A.-G. (2013). G*Power version 3.1.7 [computer software]. Uiversität Kiel, Germany. Retrieved from ▶ http://www.psycho.uni-duesseldorf.de/abteilungen/aap/gpower3/download-and-register.

Figueroa, R. L., Zeng-Treitler, Q., Kandula, S., et al. (2012). Predicting sample size required for classification performance. *BMC Medical Informatics and Decision Making, 12*, 8. ▶ https://doi.org/10.1186/1472-6947-12-8.

Gwowen Shieh, S. J., & Randles, R. H. (2006). On power and sample size determinations for the Wilcoxon–Mann–Whitney test. *Journal of Nonparametric Statistics, 18*(1), 33–43.

Guo, J.-H. (2012). Optimal sample size planning for the Wilcoxon–Mann–Whitney and van Elteren tests under cost constraints. *Journal of Applied Statistics, 39*(10), 2153–2164.

Hamilton, M. A., & Collings, B. J. (1991). Determining the appropriate sample size for nonparametric tests for location shift. *Technometrics, 3*(33), 327–337.

Israel, G. D. (1992). *Determining sample size*. University of Florida, PEOD-6. Retrieved 29 June 2019.

Julious, S. A., George, S., Machin, D., & Stephens, R. J. (1997). Sample sizes for randomized trials measuring quality of life in cancer patients. *Quality of Life Research, 1997*(6), 109–117. ▶ https://doi.org/10.1023/A:1026481815304.

Kelley, K. (2008). Sample size planning for the squared multiple correlation coefficient: Accuracy in parameter estimation via narrow confidence intervals. *Multivariate Behavioral Research, 43,* 524–555.

Kornacki, A., Bochniak, A., Kubik-Komar, A. (2017). Sample size determination in the Mann–Whitney test. *Biometrical Letters, 54*(2), 175–186. ▶ https://doi.org/10.1515/bile-2017-0010.

Lachin, J. (1977). Sample size determinations for r x c comparative trials. *Biometrics, 33*(2), 315–324. ▶ https://doi.org/10.2307/2529781.

Lei Clifton, J. B., & Clifton, D. A. (2019-03-01). Comparing different ways of calculating sample size for two independent means: A worked example. *Contemporary Clinical Trials Communications, 13,* 100309.

Matsouaka, R. A., & Betensky, R. A. (2015). Power and sample size calculations for the Wilcoxon-Mann-Whitney test in the presence of death-censored observations. *Statistics in Medicine, 34*(3), 406–431. ▶ https://doi.org/10.1002/sim.6355.

McCrum-Gardner, E. (2010). Sample size and power calculations made simple. *International Journal of Therapy and Rehabilitation, 17,* 10–14.

Moinester, M., & Gottfried, R. (2014). Sample size estimation for correlations with pre-specified confidence interval. *The Quantitative Methods for Psychology, 10*(2). ▶ https://doi.org/10.20982/tqmp.10.2.p124.

Noether, G. E. (1987). Sample size determination for some common nonparametric tests. *Journal of the American Statistical Association, 82,* 645–647.

Schlesselman, J. J. (1982). *Case-control studies: Design, conduct, analysis* (pp. 144–152). New York: Oxford University Press.

Shieh, G., Jan, S.-L., & Randles, R. H. (2007a). Power and sample size determinations for the Wilcoxon signed-rank test. *Journal of Statistical Computation and Simulation, 77*(8), 717–724. ▶ https://doi.org/10.1080/10629360600635245.

Shieh, G., Jan, S.-L., & Randles, R. (2007b). Power and sample size determinations for the Wilcoxon signed-rank test. *Journal of Statistical Computation and Simulation, 77,* 717–724. ▶ https://doi.org/10.1080/10629360600635245.

Troendle, J. F. (1999). Approximating the power of Wilcoxon's rank-sum test against shift alternatives. *Statistics in Medicine, 18,* 2763–2773. ▶ https://doi.org/10.1002/(sici)1097-0258(19991030)18:20%3c2763:aid-sim197%3e3.0.co;2-m.

Tang, Yongqiang. (2011). Size and power estimation for the wilcoxon–mann–whitney test for ordered categorical data. *Statistics in Medicine, 30*(29), 3461–3470.

van de Schoot, R., Miočević, M. (Eds.). (2020). *Small sample size solutions (Open Access): A guide for applied researchers and practitioners.* Routledge.

Zhao, Y. D., Rahardja, D., & Qu, Y (2008) Sample size calculation for the Wilcoxon–Mann–Whitney test adjusting for ties. *Statistics in Medicine, 27*(3):462–468.

# Determining Sample Size in General Linear Models

© Springer Nature Singapore Pte Ltd. 2020
J. P. Verma and P. Verma, *Determining Sample Size and Power in Research Studies*,
https://doi.org/10.1007/978-981-15-5204-5_7

- **Learning Objectives**

After going through this chapter, the readers should be able to

— Describe the steps used in G*Power software for determining sample size and power in multiple regression analysis, logistic regression, independent and repeated measures ANOVA and MANOVA experiments.

— Know the procedure in computing sample size either by estimating the effect size or by using Cohen's guidelines for effect size by using G*Power software.

— Explain the achieved power in multiple regression analysis, logistic regression, independent and repeated measures ANOVA and MANOVA experiments if sample size is fixed.

— Determine the sample size or power in different types of general linear models using G*Power software.

— Understand the effect size in different hypothesis testing experiments.

— Describe the relationship between effect size and sample size as well as between effect size and power.

**7**

## Introduction

Each statistical application requires different types of inputs for determining sample size and power in the study. In this chapter, we discuss which inputs are used for different statistical techniques such as general linear models, independent measures ANOVA, repeated measures ANOVA and MANOVA experiments. We shall discuss the procedure of deciding sample size by means of illustrations using G*Power software. Along with determining the sample size, we have discussed the procedure of computing power for a given sample size in the initial few applications. The same procedure can be adopted in other applications to compute power. We have also discussed the relationship between the sample size and effect size, where the effect size can be interpreted as the size of coefficient in a regression framework, for a given power and significance level. Also we discuss the relationship between the sample size and power when the effect size and significance levels are fixed. The readers can draw such graphs in other applications by following the same procedure.

## Linear Multiple Regression Model

In a multiple regression model, we wish to test whether a group of predictors significantly predicts an outcome variable. In other words, we wish to detect a given effect in terms of multiple correlations with some specified power and significance level in testing the significance of the multiple regression model. We shall show the procedure for determining the required sample size and power by means of the following illustration.

**Linear Multiple Regression**: Determining sample size and power.

- **Illustration 7.1**

A researcher is interested in developing a multiple regression model for estimating employees' happiness on the basis of three predictors, namely, fitness index,

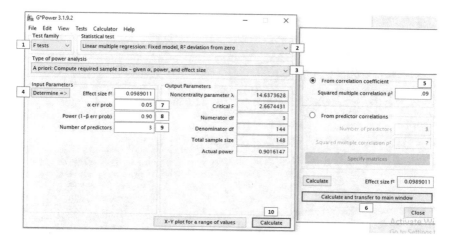

**Fig. 7.1** Steps for determining sample size in testing significance of regression model

nutritional status and hygiene. How much sample size should be required to evaluate whether these three predictors significantly predict employees' happiness and to detect a multiple correlation $(R)$ of 0.30 with power 90% at significance level 0.05?

Further, if the data is available only for 80 subjects, then what should be the power if other conditions are same?

- **Solution**

*Determining* sample size

We need to estimate the sample size that is required to test whether three predictors significantly predict employees' happiness with 90% power and having effect of multiple correlation 0.3. The following information is given in the study.

Number of predictors $= 3$
Multiple correlation $(R) = 0.3$
Significance level $= 0.05$
Power $= 0.90$

By using the above-mentioned information, we shall determine the required sample size. The below-mentioned steps in G*Power software will determine the sample size. The procedure has been shown sequentially in **Fig. 7.1**.

Step 1: Select "*F* tests".

Step 2: Select option "Linear multiple regression: Fixed Model, $R^2$ deviation from zero".

Step 3: Select "A priori" to compute the sample size.

Step 4: Click on **Determine** to compute the effect size in the adjacent window which pops up automatically.

Step 5: Enter the square of multiple correlation 0.09.

Step 6: Click on **Calculate and transfer to main window** to compute the effect size and transfer its value in the main window.

[**Remark**: Effect size $f^2$ in multiple regression is computed by the formula $f^2 = \frac{\rho^2}{1-\rho^2}$.]

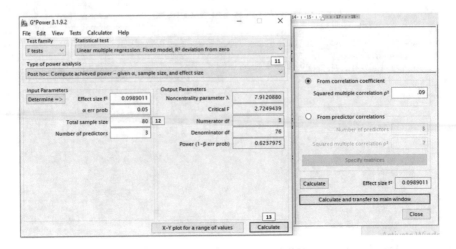

**◻ Fig. 7.2**    Steps in G-power for determining power in multiple regression analysis

7

Step 7: Enter the value of $\alpha$ as 0.05.
Step 8: Enter the value of power as 0.9.
Step 9: Enter number of predictors as 3.
Step 10: Click on **Calculate**.

After pressing **Calculate**, the sample size calculated for testing the significance of the model is 148 with achieved power 90.16%.

■  Inference
A sample of size 148 shall be required in correctly rejecting the null hypothesis that the model with three predictors does not predict employees' happiness with 90.16% power and to detect 0.3 multiple correlation with 0.05 significance level.

*Determining power if the sample size is fixed*
In the above illustration if the sample data is 80, let us see what would be the power in testing the model. In step 11, select option "Post hoc: Compute achieved power—given $\alpha$, sample size, and effect size" for determining the power as shown in ◻ Fig. 7.2. In step 12, enter the sample size as 80. All other steps would be same. Press **Calculate** to determine the power in the window.

■  Inference
With 80 as sample size, the power in correctly rejecting the null hypothesis that model does not predict employees' happiness shall be around 62.38% in the experiment as shown in ◻ Fig. 7.2.

*Relationship between sample size and effect size*
As the effect size decreases, the required sample size increases for a given number of predictors, power and significance level. The relationship can be shown by means of a

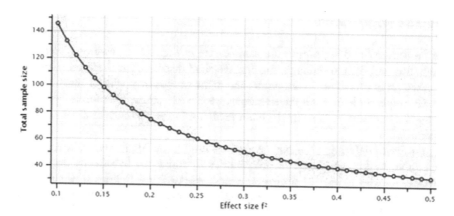

**Fig. 7.3** Relationship between the effect size and sample size in linear multiple regression for 3 predictors, 0.9 power and 0.05 significance level

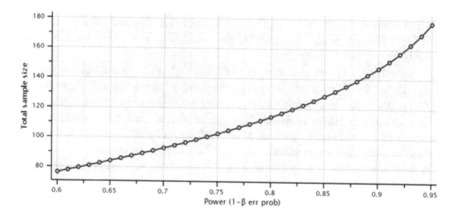

**Fig. 7.4** Relationship between the effect size and power in linear multiple regression for 3 predictors, 0.0989011 effect size ($f^2$) and 0.05 significance level

graph in ▣ Fig. 7.3 which can be obtained by clicking on **X-Y plot for a range of values** and selecting the option "Effect size $f^2$" in the screen shown in ▣ Fig. 7.1.

*Relation between sample size and power*
The power of a test is directly related to the sample size. In other words, larger power requires larger sample provided other parameters like number of predictors, effect size and significance level are fixed. The relationship between the power and sample size can be shown in ▣ Fig. 7.4 which can be obtained by clicking on **X-Y plot for a range of values** and selecting the option "Power $(1 - \beta$ err prob)" in the screen shown in ▣ Fig. 7.1.

## Logistic Regression

The procedure for determining the sample size in logistic regression differs for the continuous and dichotomous predictors. We shall discuss both these procedures, i.e. when the main predictor is continuous and dichotomous by means of Illustrations 7.2 and 7.3, respectively. In both the situations, we can determine the sample size for estimating the outcome variable by the main predictor in the presence or absence of other covariates.

In order to determine the sample size, we need to decide the effect size. It can be done either by knowing the two probabilities $P_1$ and $P_0$ or by directly deciding the effect size we wish to have in our experiment. These two probabilities are defined as

$P_1 = \text{Prob}(Y = 1|X = 1)H_1$ = Probability of occurrence of outcome variable when the main predictor variable is one standard deviation unit above its mean, and all other covariates, if exist, are assumed to set to their mean values.

$P_0 = \text{Prob}(Y = 1|X = 1)H_0$ = Probability of occurrence of outcome variable when the main predictor variable is at the mean, and all other covariates, if exist, are assumed to set to their mean values.

Let us assume that the obesity $(Y)$ is the outcome variable and calorie intake $(X)$ is the main predictor variable. The expected mean calorie intake $= 1800$ and the expected $SD = 200$. The $P_1$ can be defined as

$P_1 = \text{Prob}(Y = 1|X = 1)$ under $H_1$ = Probability of becoming obese when the main predictor variable (calorie intake) is one standard deviation unit above its mean, i.e. 2000 and all other covariates, if exist, are assumed to set to their mean values.

$P_0 = \text{Pr ob}(Y = 1|X = 1)$ under $H_0$ = Probability of becoming obese when the main predictor variable is at the mean, i.e. 1800 and all other covariates, if exist, are assumed to set to their mean values.

In determining the sample size in logistic regression, we also require the knowledge about the multiple correlation between the main predictor variable and the remaining covariates, if exist. This information can be obtained from similar studies conducted earlier or through the regression analysis. The guidelines shown in ◘ Table 7.1 can be used if estimated $R$ is unknown:

**Logistic Regression for Continuous Predictors**: Determining sample size.

- **Illustration 7.2**

An investigator is interested to test whether fat percentage influences mortality (yes 1, no 0) in patients. What should be the sample size in rejecting the null hypothesis correctly that the fat% is not associated with the probability of outcome variable (mortality) if three covariates (age, weight and calorie intake) in the study influence the outcome, with power 0.85? Use the following information in deciding the sample size:

| ◘ Table 7.1   Value of $R^2$ for different levels of association | $R$ | $R^2$ | Magnitude |
|---|---|---|---|
| | 0.2 | 0.04 | Low |
| | 0.5 | 0.25 | Medium |
| | 0.9 | 0.81 | High |

1. Type I error rate: 0.05
2. Required power: 0.85
3. Type of test: One tailed
4. $P(Y=1/X=1)$ under $H_1 =$ Probability of death $(Y=1)$ when the Fat% is $1\sigma$ unit above its mean value and all other independent variables are set to their mean values
   $= 0.30$
5. $P(Y=1/X=1)$ under $H_0 =$ Probability of death $(Y=1)$ when the Fat% is at the mean value and all other independent variables are set to their mean values
   $= 0.10$
6. $R$-square $= 0.10$ (Obtained by taking Fat% as dependent variable and other independent variables as predictors)

   **Remark**: This $R$-square represents the amount of variability in the main predictor (Fat%) that is accounted for by the other independent variables (If there are no other covariates, enter 0). The $R$-squared value can be obtained by using the regression analysis with main predictor (fat%) as a dependent variable and other covariates as independent variables. If the value of R is not known or difficult to obtain, then use thumb rule to decide its value as mentioned in ◘ Table 7.1.
7. $X$(fat%) Distribution: Assume standard normal distribution unless there is a reason to believe otherwise. Thus, take mean and standard deviation to be 0 and 1, respectively.

■ **Solution**

By using the above-mentioned information, we shall determine the required sample size in logistic regression model for testing whether Fat% (a continuous variable) is a significant predictor of mortality. The below-mentioned steps in G*Power software will determine the required sample size. The procedure has been shown sequentially in ◘ Figs. 7.5 and 7.6.

Step 1: Select "Z tests".
Step 2: Select option "Logistic regression".
Step 3: Select "A priori" to compute the sample size.
Step 4: Click on **Options** for getting the screen as shown in ◘ Fig. 7.6 for entering the effect size by two different procedures.
Step 5: Select the radio button for "Two probabilities".
Step 6: Click **OK** to come back to ◘ Fig. 7.5.
Step 7: Enter the type of test as one tail.
Step 8: Enter probability $P(Y=1/X=1)$ under $H_1$ as 0.3.
Step 9: Enter probability $P(Y=1/X=1)$ under $H_0$ as 0.1.
Step 10: Enter $\alpha$ as 0.05.
Step 11: Enter power $(1-\beta)$ as 0.85.
Step 12: Enter $R^2$ as 0.1. R is the multiple correlation between Fat% and the group of covariates (age, weight and calorie intake).

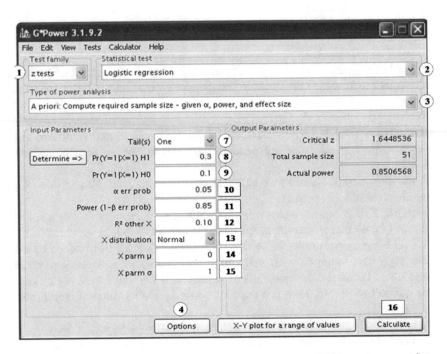

**□ Fig. 7.5**  Steps for determining sample size in logistic regression model for continuous predictors

(**Remark**: If there is only one predictor, i.e. Fat%, then enter the value of $R$-square as 0.)

Step 13: Select the distribution of Fat% ($X$) as normal by default.

Step 14: Enter the mean of $X$ (Fat%) as 0.

Step 15: Enter the SD of $X$ (Fat%) as 1.

Step 16: Click on **Calculate**.

After pressing **Calculate**, the sample size calculated for testing the significance of Fat% as a predictor of mortality is 51 for achieving 0.8507 power.

▪ **Inference**

There are 85.07% chances of correctly rejecting the null hypothesis that the main predictor variable (Fat%) is not associated with the outcome variable (mortality) with 51 subjects at 0.05 significance level in a one-tail test.

**Logistic Regression for Dichotomous Predictors**: Determining sample size.

▪ **Illustration 7.3**

A researcher is interested to investigate whether consuming tobacco affects mortality (mortality $= 1$, No mortality $= 0$) in critical patients. What should be the sample size in correctly rejecting the null hypothesis that the consuming tobacco is not associated with the mortality with or without three more covariates (gender, smoking and diet) with power 0.80? Determine the sample size using the following additional information.

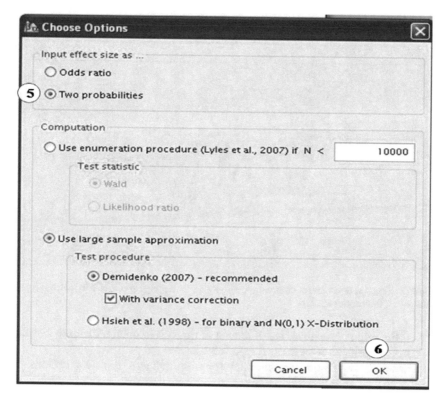

**Fig. 7.6** Steps in G*Power for choosing option for computing effect size in logistic regression

1. Level of significance: 0.05
2. Required power: 0.80
3. Type of test: two tail
4. $P(Y=1/X=1)$ under $H_1$ = Probability of death $(Y=1)$ when the subject is a tobacco user $(X=1)$
   $= 0.30$
5. $P(Y=1/X=1)$ under $H_0$ = Probability of death $(Y=1)$ when the subject is a not a tobacco user $(X=0)$
   $=0.10$
6. $R$-square $=0.25$
   (By taking tobacco as dependent variable and other three covariates; gender, smoking and diet as predictors)

### Remarks
1. If there are no other covariates, enter $R$-square as 0.
2. $R$-square represents the amount of variability in the main predictor tobacco that is accounted for by the other independent variables.
3. The $R$-square value can be obtained by regressing the main predictor tobacco on all other covariates using binary logistic regression.

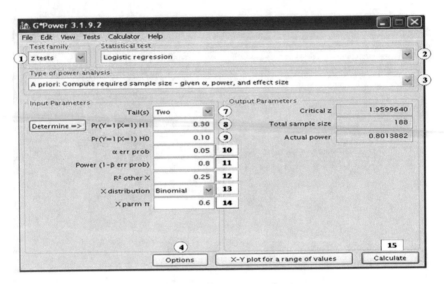

**□ Fig. 7.7**  Steps in G*Power for determining sample size in testing significance in logistic regression model for dichotomous predictor

4. If the value of $R$-square is not known or difficult to obtain, then use its value as per the guidelines mentioned in □ Table 7.1.

7. $X$(tobacco) Distribution = Binomial (because tobacco has two options only: tobacco user and non-tobacco user.)

8. $X$(tobacco) parameter = 0.6 (This is the proportion of subjects who are tobacco users.)

■ **Solution**

With the information provided above, we shall determine the required sample size in logistic regression model for testing whether a dichotomous variable (tobacco) is a significant predictor of the outcome variable (mortality). The below-mentioned steps in G*Power software will determine the required sample size. The procedure has been shown sequentially in □ Figs. 7.7 and 7.8.

Step 1: Select "Z tests".

Step 2: Select option "Logistic regression".

Step 3: Select "A priori" to compute the sample size.

Step 4: Click on **Options** to select the procedure for computing effect size in the screen as shown in □ Fig. 7.8.

Step 5: Select the radio button for "Two probabilities".

Step 6: Click **OK** to come back to the screen as shown in □ Fig. 7.7.

Step 7: Enter the type of test as two tail.

Step 8: Enter probability $P(Y=1/X=1)$ under $H_1$ as 0.3.

Step 9: Enter probability $P(Y=1/X=1)$ under $H_0$ as 0.1.

Step 10: Enter $\alpha$ as 0.05.

Step 11: Enter power $(1 - \beta)$ as 0.80.

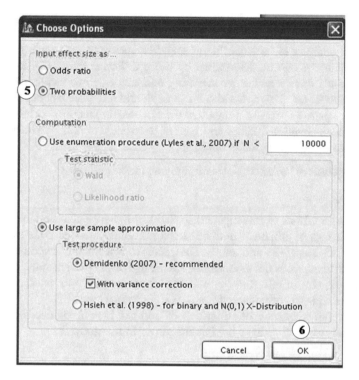

■ **Fig. 7.8** Steps in G-power for choosing option for computing effect size in logistic regression

Step 12: Enter $R^2$ as 0.25. ($R$ is the multiple correlations between tobacco consuming and the group of covariates (gender, smoking and diet). If there is only one predictor, i.e. Tobacco, then take the value of $R$-square as 0)

Step 13: Select the distribution of $X$(tobacco) as Binomial.

Step 14: Enter the parameter of $X$(tobacco) as 0.6. (This is the estimate of proportion of tobacco consumer in the population.)

Step 15: Click on **Calculate**.

After pressing **Calculate,** the sample size calculated for testing Tobacco as a significant predictor of mortality is 188 for achieving 0.8014 power.

- **Inference**

There are 80.14% chances of correctly rejecting the null hypothesis that consuming tobacco is not associated with the mortality with 188 subjects at significance level 0.05 in a two-tail test.

## Analysis of Variance

In applying analysis of variance in independent measures and repeated measures design, sample size can be determined by using G*Power software. The size of the sample depends upon whether we are using one factor or two factors with or without interaction in our design. Besides this, one should also decide about the effect (in terms of eta square or partial eta square as the case may be) required to be tested along with the power, significance level and type of test (one/two tail). We have shown all these cases in the following subsections.

**One-way Analysis of Variance**: Determining sample size *and power*.

- **Illustration 7.4**

An investigator wishes to investigate the relationship between age and pull-ups performance in college athletes. He divided the randomly selected athletes into three age groups: 18–20, 21–23 and 24–26. He wishes to test the null hypothesis that the pull-ups performance is not associated with the age of the participants. What should be the sample size in rejecting the null hypothesis correctly to detect the effect of size 0.06 (eta square) with power 0.8 at 0.05 significance level?

If the available number of athletes is 75 and the hypothesis needs to be tested with the effect size 0.08, what would be the power in the study?

- **Solution**

*Determining sample size*

Let us list the parameters given in this illustration:
$\alpha = 0.05$
Power $= 0.8$
Eta square $(\eta^2) = 0.06$
Number of groups $= 3$

Performing the following steps will determine the required sample size. The procedure has been shown sequentially in ◘ Fig. 7.9.

Step 1: Select "*F* tests" for one-way ANOVA.
Step 2: Select option "ANOVA: Fixed effects, omnibus, one-way".
Step 3: Select "A priori" to compute the sample size.
Step 4: Click on **Determine** to compute the effect size in the adjacent window, which pops up automatically.
Step 5: Select option for computing effect size from variance and choose "Direct" method.
Step 6: Enter the effect size as 0.06.

*Remark*

1. To determine sample size in one-way ANOVA, the researcher should have knowledge about the eta square $(\eta^2)$ from similar studies conducted earlier. Eta square is the estimate of the total variance in the outcome variable (pull-ups performance) accounted for by the independent variable (Age). If eta square (effect

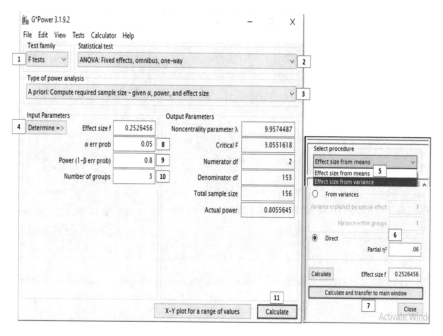

**□ Fig. 7.9** Steps in G-power for determining sample size in one-way ANOVA

| **□ Table 7.2** Guidelines for effect size | Partial eta square $\eta^2$ | Magnitude |
|---|---|---|
| | 0.02 | Small |
| | 0.06 | Medium |
| | 0.14 | Large |

size) is not known, Cohen (1988) guidelines for the effect size can be followed in this regard which is shown in □ Table 7.2.

2. G*Power uses the effect size $f$ to calculate the sample size which is obtained by using the formula $\eta^2 = \left(f^2 / \left(1 + f^2\right)\right)$.

Step 7: Click on **Calculate and transfer to main window** to compute the effect size and transfer its value in the main window.

Step 8: Enter the value of $\alpha$ as 0.05.

Step 9: Enter power as 0.8.

Step 10: Enter the number of groups as 3.

Step 11: Click on **Calculate**.

After pressing **Calculate**, the sample size and the power actually achieved are calculated for the experiment. Here, required sample size in the experiment is 156 for achieving 80.56% power.

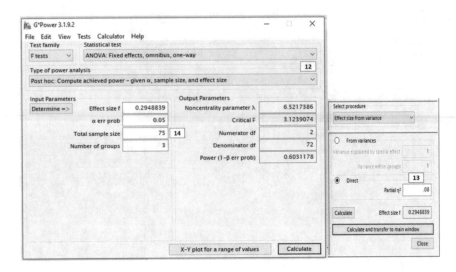

**◻ Fig. 7.10**    Steps in G-power for determining power in one-way analysis of variance technique

- **Inference**

There are 80.56% chances of correctly rejecting the null hypothesis of no difference between the pull-ups performance of athletes in different age categories with total of 156 subjects or 52 subjects per group having effect size 0.06.

*Determining Power when the sample size is fixed*

In the above study, if the sample size is 75 and the effect size is 0.08 let us see what would be the power in testing the hypothesis. In step 12, select option "Post hoc: Compute achieved power—given $\alpha$, sample size, and effect size" for determining the power as shown in ◻ Fig. 7.10. In step 13, enter the effect size (eta square) as 0.08 and in step 14 enter the sample size as 75. All other steps would be same. Press **Calculate** to determine the power in the window.

- **Inference**

With 75 as sample size, the power in correctly rejecting the null hypothesis that all the three group means of pull-ups performance are same will have 60.31% power for detecting effect size of 0.08.

**Two-way Analysis of Variance**: Determining sample size.

- **Illustration 7.5**

An investigator wishes to see the effect of gender (women vs. men) and age (9–10 years, 11–12 years and 13–14 years) on pull-ups performance. What should be the sample size for testing the significance of main and interaction effects for detecting the effect size 0.06 (partial eta square) with power 0.8 at significance level 0.05?

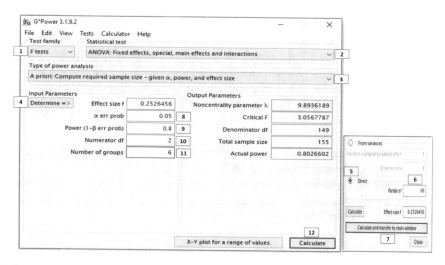

**Fig. 7.11** Steps in G-power for determining sample size for testing the main effect of age in two-way ANOVA

- **Solution**

In this illustration, we need to determine the sample size for testing the significance of main effects (gender and age) and interaction effect (gender × age). We shall first determine the required sample size for testing the main effect of 'Age' with power 0.8. Later, the same process can be replicated with minor changes to determine the sample size for the main effect of gender and interaction effect for the same power and level of significance.

In order to determine the sample size, we should have knowledge about the partial eta square ($\eta^2$) from similar studies conducted earlier. The partial eta square is the estimate of the total variance in the outcome variable (pull-ups performance) accounted for by the predictor variable (gender or age) or by the interaction (gender × age).

If partial eta square (effect size) is not known, we can follow guidelines by Cohen (1988) as shown in **Table 7.2**. To detect smaller effect, larger sample is required in comparison to detecting a larger effect for the same power. As a thumb rule, one may choose medium effect in the study.

1. *Sample size determination for testing the significance of main effect of Age*

In this illustration, the following information is given:
$\alpha = 0.05$
Power $= 0.8$
Partial eta square $(\eta^2) = 0.06$

Performing the following steps will determine the required sample size. The procedure has been shown sequentially in **Fig. 7.11**.
Step 1: Select "$F$ tests" for two-way ANOVA.
Step 2: Select option "ANOVA: Fixed effects, special, main effects and interactions".
Step 3: Select "A priori" to compute the sample size.

Step 4: Click on **Determine** to compute the effect size in the adjacent window which pops up automatically.

Step 5: Select "Direct" option for computing effect size based on partial eta square $\eta^2$.

Step 6: Enter the value of $\eta^2$ as 0.06 for the main effect age.

Step 7: Click on **Calculate and transfer to main window** to compute the effect size and transfer its value in the main window.

Step 8: Enter $\alpha$ as 0.05.

Step 9: Enter power as 0.8.

Step 10: Enter the degrees of freedom for age as 2. Since there are three age categories, hence, degrees of freedom for age is $3 - 1 = 2$.

Step 11: Enter total number of groups in the experiment as 6. Since there are three age categories and two gender groups, hence, total number of groups is $3 \times 2 = 6$.

Step 12: Click on **Calculate**.

After pressing **Calculate,** the sample size and actual power achieved are calculated for the experiment. Here, required sample size in the experiment is 155 for achieving 80.27% power in testing the null hypothesis for the main effect of Age. Dividing 155 into six groups results in 25.7($\approx$26). Thus, 26 subjects are required in each group resulting in 52 students in each age category (women and men combined).

- **Inference**

There are 80.27% chances of correctly rejecting the null hypothesis of no difference between the mean pull-ups performance of the subjects in different age categories with total of 156 students or 26 students per group having the effect of 0.06 eta square.

2. *Sample size determination for testing the significance of main effect of **Gender***

In order to test the effect of gender on pull-ups performance with 0.8 power, the same process as discussed above shall be followed except in Step 10 where you need to enter the degrees of freedom of gender as 1 because there are only two gender groups. After choosing this option, the screen shall look like $\blacksquare$ Fig. 7.12.

Here, required sample size in the experiment is 125 for achieving 0.8 power in testing the effect of gender. Dividing 125 into six groups results in 20.8($\approx$21). Thus, 21 subjects are required in each group resulting in 63 students in each gender category.

- **Inference**

There are 80% chances of correctly rejecting the null hypothesis of no difference between the pull-ups performance of women and men with total of 126 students or 63 students per gender category having effect size of 0.06.

3. *Sample size determination for testing the interaction effect of **Age × Gender** with power 0.8*

Since the degrees of freedom for interaction is 2 which is same as that of the age factor, hence, same number of students as determined in section (a) for the main effect of

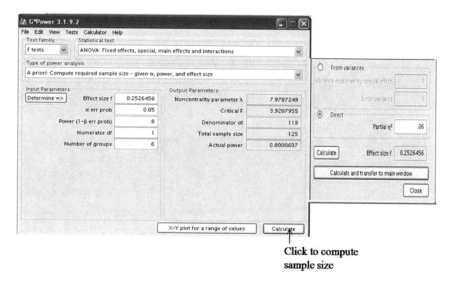

Click to compute
sample size

**□ Fig. 7.12** Steps in G-power for determining sample size in testing main effect of gender in a two-way ANOVA

age above shall be required for testing the interaction effect with 0.8 power and effect size 0.06.

**Repeated Measures ANOVA between Factors**: Determining sample size.

- **Illustration 7.6**

An exercise scientist wishes to conduct an exercise efficacy test to see its effectiveness on VO2max in 2-week duration. He wishes to conduct the experiment on 30 women subjects. Fifteen subjects in the treatment group receive the exercise, whereas 15 subjects in the control group get placebo. The group of subjects getting placebo treatment acts a control group. All subjects are tested for VO2 max before start of the test (pre-test) after 2 weeks (post-test) and 4 weeks (follow-ups). What should be the estimated sample size for detecting a significant difference in exercise efficacy results between the treatments and control groups with 0.8 power and Type I error rate as 0.05? Similar studies conducted earlier reported standard deviation of VO2 max as 2.5, and means of the exercise and control groups are expected to have as 51 and 49, respectively.

- **Solution**

In this illustration, the following information is provided:
Estimated SD of VO2 max in the population $= 2.5$
Mean of the experimental group $= 51$
Mean of the control group $= 49$
Sample size $= 15$ $(n_1 = n_2)$
$\alpha = 0.05$
Power $= 0.8$

7

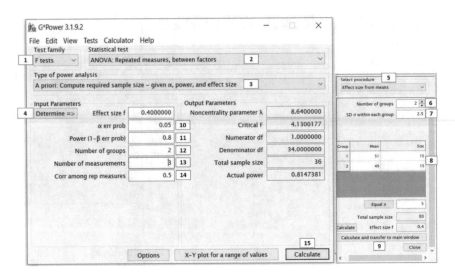

■ **Fig. 7.13** Steps in G-power for determining sample size in testing the effect of between factors in repeated measures ANOVA

Number of groups = 2(Treatment and control)
Number of measurements = 3(There are three time periods)
Correlation among repeated measures = 0.5.

**Remark** In repeated measures design the dependent variable in repeated measures should have moderate to high correlation. By default, one can take it as 0.5.

Performing the following steps will determine the required sample size. The procedure has been shown sequentially in ■ Fig. 7.13.

Step 1: Select "*F* tests" for repeated ANOVA.
Step 2: Select option "ANOVA: Repeated measures, between factors".
Step 3: Select "A priori" to compute the sample size.
Step 4: Click on **Determine** to compute the effect size in the adjacent window which pops up automatically.
Step 5: Select "Effect size from means" option for computing effect size based on means of groups.
Step 6: Enter number of groups as 2 (treatment and control).
Step 7: Enter the population standard deviation as 2.5.
Step 8: Enter expected means and sample size in each group. Here, means of both the groups are 51 and 49, and expected sample size in each group is 15.
   **Remark** In case the information about means and standard deviation are not known, one can use the guidelines for the effect size 'f' (as shown in ■ Table 7.3) directly in the window. Usually medium effect size is taken.
Step 9: Click on **Calculate and transfer to main window** to compute the effect size and transfer its value in the main window.
Step 10: Enter α as 0.05.
Step 11: Enter power as 0.8.

| ◻ Table 7.3  Effect size (*f*) convention | Effect size | Magnitude |
|---|---|---|
| | 0.10 | Small |
| | 0.25 | Medium |
| | 0.40 | Large |

Step 12: Enter the number of groups as 2 (since there are two groups treatment and control).

Step 13: Enter number of measurements as 3 (since there are three repeated measures).

Step 14: Enter correlation as 0.5. One should usually take the default correlation as 0.5 in the absence of this information from the similar studies conducted previously.

Step 15: Click on **Calculate**.

After pressing **Calculate**, the sample size and power actually achieved are obtained. Here, required sample size in the experiment is 36 for achieving 0.8147 power.

- **Inference**

There are 81.47% chances of correctly rejecting the null hypothesis of no difference between experimental and control groups in VO2 max with a total of 36 subjects in the experiment, 18 in each group.

*Remark* In testing between the groups when the repeated measurements are obtained, usually scores on the outcome variable across all repeated measures are added for each participant, and the sum so obtained is divided by the square root of the number of repeated measures. For this example, each subject's score would be calculated by using the formula (test1 + test2 + test3)/√3. The software then compares these scores using the grouping variable (i.e. treatment vs. control).

**Repeated Measures ANOVA within Factor**: Determining sample size.

- **Illustration 7.7**

A medical researcher wishes to conduct an experiment with 20 subjects for testing the effectiveness of a drug in enhancing the WBC blood count across the time. The WBC counts of the subjects shall be tested after 1, 2 and 3 weeks. What should be the sample size to detect a significant difference in WBC counts across time (1, 2 and 3 weeks) with effect size 0.08 and power 0.9 at significance level 0.05?

- **Solution**

To find the sample size in testing the effectiveness of a drug across the time, the following information is given:

Partial eta square ($\eta^2$) = 0.08

$\alpha = 0.05$

Power = 0.9

Number of groups = 1

Number of measurements = 3(There are three times of testing; 1, 2 and 3 weeks)

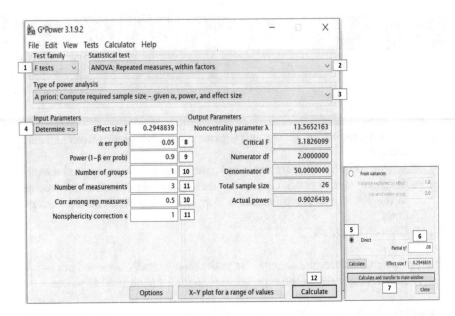

**◘ Fig. 7.14**  Steps in G-power for determining sample size in testing within group effect in repeated measures ANOVA

Correlation among repeated measures = 0.5

[**Remark**: In repeated measures design, the dependent variable in repeated measures should have moderate to high correlation. By default, one can take it as 0.5.]

Non-sphericity correction = 1

[**Remark**: We have assumed that there is no violation of sphericity. In case sphericity assumption is violated, one should use the value of ε which can have values in between 0 and 1. The procedure of testing sphericity assumption and computing ε can be seen in "Repeated Measures Design for Empirical Researchers" (Verma 2015).]

Performing the following steps will determine the required sample size in repeated measures ANOVA within factor. The procedure has been shown sequentially in ◘ Fig. 7.14.

Step 1: Select "*F* tests" as a family of test.

Step 2: Select option "ANOVA: Repeated measures, within factors".

Step 3: Select "A priori" to compute the sample size.

Step 4: Click on **Determine** to compute the effect size in the adjacent window, which pops up automatically.

Step 5: Select "Direct" option for computing effect size (*f*) based on partial eta square $\eta^2$.

Step 6: Enter partial eta square $\eta^2$ as 0.08 for medium effect.

Step 7: Click on **Calculate and transfer to main window** to compute effect size and transfer its value in the main window.

Step 8: Enter $\alpha$ as 0.05.

Step 9: Enter power as 0.90.

Step 10: Enter the number of groups as 1.

Step 11: Enter number of measurements as 3 (Since there are three levels of within variables).

Step 12: Enter correlation as 0.5.

Step 13: Enter non-sphericity correction ε as 1. This is so because we have assumed that sphericity assumption has not been violated.

Step 14: Click on *Calculate*.

After pressing *Calculate*, the sample size calculated for testing the effect of within factor is 26 for achieving 0.9026 power.

■ **Inference**

There are 90.26% chances of correctly rejecting the null hypothesis of no difference in WBC counts across the time with 26 subjects and having effect size 0.08 and significance level 0.05.

**Repeated Measures ANOVA for testing interaction:** Determining sample size.

■ **Illustration 7.8**

An exercise scientist wishes to conduct an experiment with 30 subjects (15 men and 15 women) for testing effectiveness of an exercise programme on VO2 max. Subjects' VO2 max shall be measured at zero week, 2 weeks and 4 weeks. Duration is within-subject variable, and gender is between-subject variable. What should be the sample size in correctly rejecting the null hypothesis of no interaction between duration and gender for having effect size 0.06 (partial eta square) with 0.85 power at significance level 0.05?

■ **Solution**

Here, we need to determine the sample size for testing the null hypothesis of no interaction between gender and duration on VO2 max with power 0.85 and effect size 0.06. In order to determine the sample size, we should have knowledge about the partial eta square $\eta^2$ from the earlier studies. The partial eta square is the estimate of the total variance in the outcome variable (VO2 max) accounted for by a predictor variable (in this case, it is the interaction between duration and gender).

If partial eta square (effect size) is not known, we can use 0.02 for small effect, 0.06 for medium effect and 0.14 for large effect. By default one may use medium effect. Thus, we have the following information:

Partial Eta square ($\eta^2$) = 0.06

$\alpha = 0.05$

Power = 0.85

Number of groups = 2(duration and gender)

Number of measurements = 3(There are three time periods zero, 2 and 4 weeks)

Correlation among repeated measures = 0.5 (correlation among columns of repeated measures)

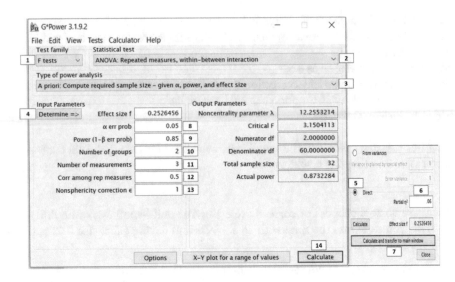

**◘ Fig. 7.15**   Steps in G-power for determining sample size in testing interaction between a between-subject variable and within-subject variable in repeated measures ANOVA

[**Remark**: In repeated measures design, the dependent variable in repeated measures should have moderate to high correlation. By default, one can take it as 0.5.]

Non-sphericity correction $= 1$

[**Remark**: We have assumed that there is no violation of sphericity. In case sphericity is violated, one should use the value of $\varepsilon$ which can have values in between 0 and 1. The procedure of testing sphericity assumption and computing $\varepsilon$ can be seen in "Repeated Measures Design for Empirical Researchers" (Verma 2015).]

Performing the following steps will determine the required sample size. The procedure has been shown sequentially in ◘ Fig. 7.15.

Step 1: Select "F tests" in test family.

Step 2: Select option "ANOVA: Repeated measures, within-between interaction".

Step 3: Select "A priori" to compute the sample size.

Step 4: Click on **Determine** to compute the effect size in the adjacent window, which pops up automatically.

Step 5: Select "Direct" option for computing effect size ($f$) based on partial eta square $\eta^2$.

Step 6: Enter partial eta square $\eta^2$ as 0.06 for medium effect.

Step 7: Click on **Calculate and transfer to main window** to compute effect size and transfer its value in the main window.

Step 8: Enter $\alpha$ as 0.05.

Step 9: Enter power as 0.85.

Step 10: Enter the number of groups as 2 (duration and gender).

Step 11: Enter number of measurements as 3 (Since there are three levels of within variables).

Step 12: Enter correlation as 0.5. In the absence of the knowledge about correlation from earlier studies, one should usually take default correlation as 0.5. This is the correlation in dependent variables in different repeated measures.

Step 13: Enter non-sphericity correction $\varepsilon$ as 1. This is so because we have assumed that sphericity assumption has not been violated.

Step 14: Click on **Calculate**.

After pressing **Calculate**, the sample size calculated for testing the effect of interaction is 32 for achieving 0.8732 power.

- **Inference**

There are 87.32% chances of correctly rejecting the null hypothesis of no significant effect of the interaction with 16 men and 16 women with a total of 32 subjects having effect size 0.06 and significance level 0.05.

**MANOVA For Testing the Significance of Global Effect**: Determining sample size.

- **Illustration 7.9**

An exercise scientist is interested to compare three different intensities of aerobic exercise in a 4-week experiment on health-related fitness. The health-related fitness consists of five parameters, namely, cardiorespiratory endurance, muscular endurance, muscular strength, flexibility and fat%. What sample size should be taken in testing the null hypothesis of no effect of aerobic exercise on health-related fitness with power 0.9 at significance level 0.05 for estimated value of Pillai's V as 0.5? Pillai's V is estimated by running a MANOVA on preliminary data and if such data is not available you can generate it by using random sampling based on estimates of mean and standard deviation.

- **Solution**

We need to estimate the sample size for testing the effect of aerobic exercise on the health-related parameters as a whole. In other words, it is required to test the null hypothesis which states that all the three treatments (three different intensities of aerobic exercise) are equally effective. Following parameters are given in this illustration:

$\alpha = 0.05$
Power $= 0.90$
Pillai's V $= 0.5$
Number of treatment groups $= 3$
Number of response variables $= 5$

Performing the following steps will determine the required sample size. The procedure has been shown sequentially in ◻ Figs. 7.16 and 7.17.

Step 1: Select "F tests".

Step 2: Select option "MANOVA: Global effects".

Step 3: Select "A priori" to compute the sample size.

Step 4: Click on **Options** to choose multivariate test. This will take you to the screen shown in ◻ Fig. 7.17.

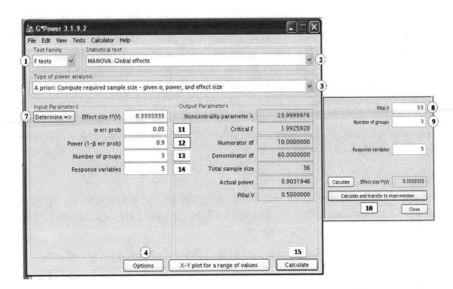

**Fig. 7.16** Steps in G*Power for determining sample size in testing the significance of global effect in MANOVA experiment

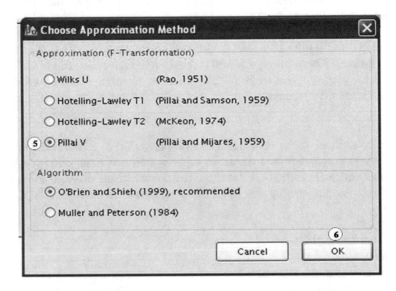

**Fig. 7.17** Option for selecting test statistic in MANOVA experiment

Step 5: Choose "Pillai's V" test. You can choose another test like Wilks' U or Hotelling as well.

Step 6: Click **OK** for test selection and getting back to **Fig. 7.16**.

Step 7: Click on **Determine** to compute the effect size in the adjacent window which pops up automatically.

Step 8: Enter the value of estimated Pillai's V as 0.5.

**Remark**: In case the value of Pillai's V is not known, then one can directly enter the effect size $f^2(V)$ in the screen shown in ◘ Fig. 7.16 by using the guidelines given in ◘ Table 7.4.

Step 9: Enter the number of groups as 3. This is the number of treatments we have.

Step 10: Click on **Calculate and transfer to main window** to compute the effect size and transfer its value in the main window.

Step 11: Enter the value of $\alpha$ as 0.05.

Step 12: Enter power as 0.9.

Step 13: Enter the number of groups as 3 in the illustration.

Step 14: Enter the number of response variables as 5. In this illustration, there are five dependent variables.

Step 15: Click on **Calculate**.

Once you click on **Calculate**, the sample size is calculated as shown in ◘ Fig. 7.16. Here, the required sample size is 36 for achieving 0.9032 power.

- **Inference**

There are 90.32% chances of correctly rejecting the null hypothesis of no difference between the performances on health-related parameters in different treatment groups with total of 36 subjects (12 in each group).

**MANOVA for Testing Significance of Interaction Effect:** Determining sample size.

- **Illustration 7.10**

Consider a situation in which the effects of three different intensities of aerobic exercise are to be compared on improving fitness (consists of cardiorespiratory endurance, muscular strength and flexibility) in men and women both. It is assumed that the subjects receive one of the three intensities of aerobic exercise. What sample size should be taken in correctly rejecting the null hypothesis of no interaction effect of aerobic exercise and gender on fitness with power 0.9 at significance level 0.05 for detecting the effect size represented by Pillai's V as 0.25? Pillai's V is estimated by running a MANOVA on preliminary data.

- **Solution**

In this illustration, the following information is provided:

$\alpha = 0.05$

Power $= 0.90$

Pillai's V $= 0.25$

Number of Independent variables $= 2$ (Aerobic exercise and Gender)

Total Number of groups $= 6$ (3 Exercise $\times$ 2 Gender)

Number of response variables $= 3$ (three fitness variables)

Following the below-mentioned steps will determine the required sample size. The procedure has been shown sequentially in ◘ Figs. 7.18 and 7.19.

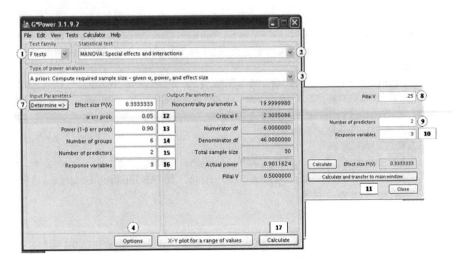

**7**

■ **Fig. 7.18** Steps in G-power for determining sample size in testing the significance of interaction effect in MANOVA experiment

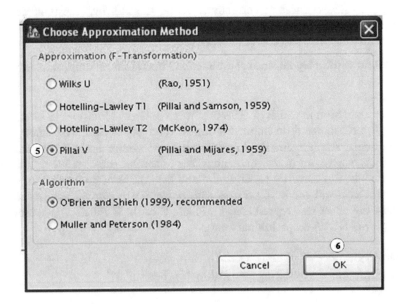

■ **Fig. 7.19** Optionfor selecting test statistic in MANOVA experiment

Step 1: Select "*F* tests".
Step 2: Select option "MANOVA: Special effects and interactions".
Step 3: Select "A priori" to compute the sample size.
Step 4: Click on **Options** to choose multivariate test. This will take you to the screen shown in ■ Fig. 7.19.
Step 5: Choose "Pillai's V" test. You can choose another test like Wilks' U or Hotelling as well.

| ▣ **Table 7.4** Guidelines for effect size | Effect size ($f^2(V)$) | Magnitude |
|---|---|---|
| | 0.01 | Small |
| | 0.06 | Medium |
| | 0.16 | Large |

Step 6: Click **OK** for test selection and getting back to the screen shown in ▣ Fig. 7.18.

Step 7: Click on **Determine** to compute the effect size in the adjacent window which pops up automatically.

Step 8: Enter the value of estimated Pillai's V as 0.25.

> **Remark**: In case the value of Pillai's V is not known, then one can directly enter the effect size $f^2(V)$ in the screen shown in ▣ Fig. 7.18 by using the guidelines given in ▣ Table 7.4.

Step 9: Enter the number of predictors as 2 as there are two predictors "Aerobic exercise" and "Gender".

Step 10: Enter response variables as 3. There are three dependent variables (cardiorespiratory endurance, muscular strength and flexibility).

Step 11: Click on **Calculate and transfer to main window** to compute the effect size and transfer its value in the main window.

Step 12: Enter the value of $\alpha$ as 0.05.

Step 13: Enter power as 0.9.

Step 14: Enter the number of groups as 6 (3 Exercise × 2 Gender).

Step 15: Enter the number of predictors as 2.

Step 16: Enter number of response variables as 3.

Step 17: Click on **Calculate**.

Click on **Calculate** to determine the sample size as shown in ▣ Fig. 7.18. Here, the required sample size is 30 for achieving power 0.9012.

- ▪ Inference

There are 90.12% chances of correctly rejecting the null hypothesis of no interaction between gender and aerobic exercise with total of 30 subjects (5 in each group).

## Summary

By using the G*Power software, required sample size can be calculated in applications like multiple linear regression, logistic regression, different independent and repeated measures, ANOVA and MANOVA experiments. Cohen has suggested guidelines for low, medium and high effect size as 0.2, 0.5 and 0.8, respectively, for general experiments. However, guidelines for the effect size differ in multiple regression, logistics regression, ANOVA and MANOVA experiments which have been discussed in different applications in this chapter. If the sample size is fixed, then the achieved power can be computed in these applications.

- **Exercises**
1. What do you mean by effect size in one-way and two-way ANOVA experiments? Explain by means of examples.
2. Why effect size is computed differently in different applications. Discuss by means of examples.
3. Describe the effect size computations by the indices $f^2$ in multiple regression, $f$ and partial eta square $(\eta^2)$ in ANOVA, $f^2(V)$ in MANOVA.
4. What is the difference between eta square and partial eta square? Explain by means of an example.
5. What parameters did you require to find the sample size in logistic regression for continuous predictors and dichotomous predictors? Explain their meaning as well.
6. A medical scientist wishes to develop a multiple regression model for estimating systolic pressure on the basis of four predictors, namely age, BMI, hours of physical activity and hours of sleep. What should be the sample size in correctly rejecting the null hypothesis that the four predictors do not predict systolic pressure and detect $R^2$ of value 0.25 in the model with power 0.8 at significance level 0.05? Further, if sample of size 60 is available then what should be the power if other conditions are same?
7. A medical researcher is interested to identify whether a tumour is malignant (1) or benign (0). Different imaging techniques will be used to extract three features of tumours: tumour size, affected body area and tumour age. Determine the sample size in correctly rejecting the null hypothesis that the tumour size is not associated with the probability of outcome variable (malignancy) with 0.8 power if two covariates (affected body area and tumour age) in the study influence the outcome. Use the following information in deciding the sample size:
    1. Type I error rate: 0.05
    2. Required power: 0.80
    3. Type of test: One tailed
    4. $P(Y=1/X=1)$ under $H_1$ = Probability of having tumour $(Y=1)$ when the tumour size is $1\sigma$ unit above its mean value and all other independent variables are set to their mean values
       $=0.25$
    5. $P(Y=1/X=1)$ under $H_0$ = Probability of having tumour $(Y=1)$ when the tumour size is at the mean value and all other independent variables are set to their mean values
       $=0.15$
    6. R-square$=0.35$ (Obtained by taking tumour size as dependent variable and other independent variables as predictors)
    7. $X$(tumour size) Distribution: Assume standard normal distribution. Thus, take mean and standard deviation to be 0 and 1, respectively.
8. A management consultant is interested in investigating whether gender affects buying behaviour of perfume (buying$=1$, not buying$=0$). What sample size he should take in correctly rejecting the null hypothesis that the gender is not associated with buying of perfume with or without two more covariates (Age category, SES category) with power 0.90? Use the following information in deciding the sample information:

1. Level of significance: 0.05
2. Required power: 0.90
3. Type of test: one tail
4. $P(Y=1/X=1)$ under $H_1$ = Probability of buying $(Y=1)$ when the subject is female $(X=1)$
   = 0.40
5. $P(Y=1/X=1)$ under $H_0$ = Probability of buying $(Y=1)$ when the subject is male $(X=0)$
   = 0.20
6. $R$-square = 0.35
7. $X$(Gender) Distribution = Binomial (because gender has two options: male and female.)
8. $X$(Gender) parameter = 0.65 (This is the proportion of subjects who are female.)
9. An educational researcher is interested in investigating the effect of duration of class on learning efficiency. Out of three randomly selected groups of subjects, first will undergo for 1-h class and second will undergo for 1.30 h class, whereas the third group will be given 2-h class. What sample size is required to detect the effect size 0.08 (eta square) with power 0.9 at significance level 0.01?
   If the number of subjects is 165 and the hypothesis needs to be tested for the effect size 0.06, what would be the power in the study?
10. An exercise scientist wishes to see the effect of age (young vs. old) and background music (instrumental, classical and jazz) on mood of the participants. What sample size he should take for testing the significance of main and interaction effects for detecting the effect size 0.08 (partial eta square) with power 0.9 at significance level 0.05?
11. A psychologist desires to investigate the effect of pranayama on concentration. One group of subjects will be given pranayama, whereas the second group of subjects will act as a control and will be given placebo treatment. All subjects in experimental and control groups will be tested for concentration before start of the experiment, after 3 weeks and 6 weeks. What sample size he should take to correctly reject the null hypothesis of no difference between the treatment and control groups 80% of time having effect size 0.06 (eta square) at significance level 0.05 in a one-tailed test?
12. A pharmacologist is interested to test the effectiveness of an herbal drug on haemoglobin among the patients. After administering the drug daily, he plans to test the subjects for haemoglobin after 2, 4 and 6 weeks. What sample size he should take to detect a significant difference in haemoglobin level across time (2, 4 and 6 weeks) with effect size 0.06 and power 0.8 at significance level 0.05? It is assumed that there is no violation of sphericity in the data and correlation among repeated measures is 0.5.
13. A researcher plans to conduct an experiment for testing the effectiveness of a spiritual programme on happiness among men and women. The happiness of the subjects will be measured in zero, 2, 4 and 6 weeks. Time is within-subject variable, and gender is between-subject variable. What should be the sample size to detect a significant interaction between the time and gender with effect size (partial eta square) 0.07 and power 0.80 at significance level 0.01? It is assumed that there is no violation of sphericity in the data and correlation among repeated measures is 0.5.

14. A psychological researcher wishes to investigate the effect of weather conditions (cold, hot and humid) on subject's frustration discomfort (measured with four dimensions: emotional tolerance, entitlement, discomfort tolerance and achievement) by using the MANOVA analysis. What sample size should be taken to detect the effect of weather condition on frustration discomfort with 0.8 power and significance level 0.05 if estimated Pillai's V is 0.3?

15. A researcher is interested to conduct a study to investigate the effect of job type (banking, insurance and telecommunication) on quality of life (consists of four variables: creativity, learning, well-being and memory) among men and women executives. What sample size he should take to detect an interaction between job type and gender with 0.9 power if $\alpha = 0.05$ and estimated Pillai's V is 0.2?

- ■ Answer
  6. (a)  Total sample size $= 41$
     (b)  For $n = 60$, power of correctly rejecting the null hypothesis would be 94.73%
  7. $n = 190$
  8. $n = 383$
  9. (a)  Total of 207 subjects or 69 subjects per group required
     (b)  Power $= 62.85\%$)
  10. (a)  Sample size required for testing the significance of main effect of age is 21 subjects per group.
      (b)  Sample size required for testing the significance of main effect of music is 25 subjects per group.
      (c)  Sample size required for testing the interaction effect of age $\times$ music is 25 subjects per group.
  11. Required sample size in experimental group is 42 and in control group is also 42.
  12. Required sample size is 27.
  13. 14 men and 14 men would be required in the study.
  14. The sample size in each of the three groups should be 16.
  15. Total sample required is 43 or seven subjects in each of the six groups.

## Bibliography

Beaujean, A. A. (2014). Sample size determination for regression models using Monte Carlo methods in R. *Practical Assessment, Research & Evaluation, 19*(12). Retrieved from ► http://pareonline.net/getvn.asp/v=19&n=12.

Bujang, M. A., Sa'at, N., & Bakar, T. M. I. T. A. (2017). Determination of minimum sample size requirement for multiple linear regression and analysis of covariance based on experimental and non-experimental studies. *Epidemiology Biostatistics and Public Health, 14*(3).

Bush, S. (2014). Sample size determination for logistic regression: A simulation study. *Communication in Statistics- Simulation and Computation, 44*. 10.1080/03610918.2013.777458.

Chandrasekaran, V., Gopal, G., & Basilea, W. (2012). Sample size determination for repeated measurements using three possible methods of analysis, POST, CHANGE and ANCOVA, under various covariance structures for two groups. *Asian Journal of Applied Sciences, 5*, 371–383.

Cohen J., (1988). *Statistical power analysis for the behavioral science* (2nd ed., pp. 284–288) Hillsdale: Lawrence Erlbaum Associates.

Dupont, W. D., & Plummer, W. D. (1998). Power and sample size calculations for studies involving linear regression. *Controlled Clinical Trials, 19,* 589–601.

Faul, F., Erdfelder, E., Buchner, A., & Lang, A. (2009). Statistical power analyses using G*Power 3.1: Tests for correlation and regression analyses. *Behavior Research Methods, 41*(4), 1149–1160. ▶ https://doi.org/10.3758/brm.41.4.1149.

Frane, Andrew. (2015). Power and type I error control for univariate comparisons in multivariate two-group designs. *Multivariate Behavioral Research, 50*(2), 233–247. ▶ https://doi.org/10.1080/0 0273171.2014.968836.PMID26609880.

Garson, G. D. *Multivariate GLM, MANOVA, and MANCOVA.* Retrieved March 22, 2019.

Guo, Y., Logan, H. L., Glueck, D. H., et al. (2013). Selecting a sample size for studies with repeated measures. *BMC Medical Research Methodology, 13,* 100. ▶ https://doi.org/10.1186/1471-2288-13-100.

Hsieh, F. Y., Bloch, D. A., & Larsen, M. D. (1998). A simple method of sample size calculation for linear and logistic regression. *Statistics in Medicine, 17,* 1623–1634.

James, A. H, (2016). Simple and multiple linear regression: Sample size considerations. *Journal of Clinical Epidemiology, 79,* 112–119.

Jan, S., & Shieh, G. (2019). Sample size calculations for model validation in linear regression analysis. *BMC Medical Research Methodology, 19,* 54. ▶ https://doi.org/10.1186/s12874-019-0697-9.

Kang, H. (2015). Sample size determination for repeated measures design using G Power software. *Anesthesia and Pain Medicine, 10.* 6–15. ▶ https://doi.org/10.17085/apm.2015.https://doi.org/10.1.6.

Kelley, K., & Maxwell, S. E. (2003). Sample size for multiple regression: Obtaining regression coefficients that are accurate.

Ken, K., Scott, E. M. (2003). Sample size for multiple regression: Obtaining regression coefficients that are accurate, not simply significant. *Psychological Methods, 8*(3), 305–321.

Kim, H. Y. (2016). Statistical notes for clinical researchers: Sample size calculation 3. Comparison of several means using one-way ANOVA. *Restorative Dentistry & Endodontics, 41*(3), 231–234. ▶ https://doi.org/10.5395/rde.2016.41.3.231.

Kim, S., Heath, E., & Heilbrun, L. (2017). Sample size determination for logistic regression on a logit-normal distribution. *Statistical Methods in Medical Research, 26*(3), 1237–1247. ▶ https://doi.org/10.1177/0962280215572407.

Muller, K. E., Lavange, L. M., Ramey, S. L., & Ramey, C. T. (1992). Power calculations for general linear multivariate models including repeated measures applications. *Journal of American Statistical Association, 87,* 1209–1226.

Self, S. G., & Mauritsen, R. H. (1988). Power/sample size calculations for generalized linear models. *Biometrics, 44,* 79–86.

Stevens, J. P. (2002). *Applied multivariate statistics for the social sciences.* Mahwah, NJ: Lawrence Erblaum.

Taylor, A. (2011) JMASM31: MANOVA procedure for power calculations (SPSS). *Journal of Modern Applied Statistical Methods, 10*(2), Article 33. ▶ https://doi.org/10.22237/jmasm/1320121920.

Warne, R. T. (2014). A primer on multivariate analysis of variance (MANOVA) for behavioral scientists. *Practical Assessment, Research & Evaluation, 19*(17), 1–10.

Whittemore, A. (1981). Sample size for logistic regression with small response probability. *Journal of the American Statistical Association, 76,* 27–32.

Wolfe, R., & Carlin, J. B. (1999). Sample-size calculation for a log-transformed outcome measure. *Controlled Clinical Trials, 20,* 547–554.

# Supplementary Information

## Appendix – 2

© Springer Nature Singapore Pte Ltd. 2020
J. P. Verma and P. Verma, *Determining Sample Size and Power in Research Studies*,
https://doi.org/10.1007/978-981-15-5204-5

# Appendix

See ■ Tables A.1 and A.2.

**Table A.1** The normal curve area between the mean and a given z value

| Z | α for one-tailed test | | | | | | | | | |
|---|---|---|---|---|---|---|---|---|---|---|
| | 0.00 | 0.01 | 0.02 | 0.03 | 0.04 | 0.05 | 0.06 | 0.07 | 0.08 | 0.09 |
| 0.0 | 0.0000 | 0.0040 | 0.0080 | 0.0120 | 0.0160 | 0.0199 | 0.0239 | 0.0279 | 0.0319 | 0.0359 |
| 0.1 | 0.0398 | 0.0438 | 0.0478 | 0.0517 | 0.0557 | 0.0596 | 0.0636 | 0.0675 | 0.0714 | 0.0753 |
| 0.2 | 0.0793 | 0.0832 | 0.0871 | 0.0910 | 0.0948 | 0.0987 | 0.1026 | 0.1064 | 0.1103 | 0.1141 |
| 0.3 | 0.1179 | 0.1217 | 0.1255 | 0.1293 | 0.1331 | 0.1368 | 0.1406 | 0.1443 | 0.1480 | 0.1517 |
| 0.4 | 0.1554 | 0.1591 | 0.1628 | 0.1664 | 0.1700 | 0.1736 | 0.1772 | 0.1808 | 0.1844 | 0.1879 |
| 0.5 | 0.1915 | 0.1950 | 0.1985 | 0.2019 | 0.2054 | 0.2088 | 0.2123 | 0.2157 | 0.2190 | 0.2224 |
| 0.6 | 0.2257 | 0.2291 | 0.2324 | 0.2357 | 0.2389 | 0.2422 | 0.2454 | 0.2486 | 0.2517 | 0.2549 |
| 0.7 | 0.2580 | 0.2611 | 0.2642 | 0.2673 | 0.2704 | 0.2734 | 0.2764 | 0.2794 | 0.2823 | 0.2852 |
| 0.8 | 0.2881 | 0.2910 | 0.2939 | 0.2967 | 0.2995 | 0.3023 | 0.3051 | 0.3078 | 0.3106 | 0.3133 |
| 0.9 | 0.3159 | 0.3186 | 0.3212 | 0.3238 | 0.3264 | 0.3289 | 0.3315 | 0.3340 | 0.3365 | 0.3389 |
| 1.0 | 0.3413 | 0.3438 | 0.3461 | 0.3485 | 0.3508 | 0.3531 | 0.3554 | 0.3577 | 0.3599 | 0.3621 |
| 1.1 | 0.3643 | 0.3665 | 0.3686 | 0.3708 | 0.3729 | 0.3749 | 0.3770 | 0.3790 | 0.3810 | 0.3830 |
| 1.2 | 0.3849 | 0.3869 | 0.3888 | 0.3907 | 0.3925 | 0.3944 | 0.3962 | 0.3980 | 0.3997 | 0.4015 |
| 1.3 | 0.4032 | 0.4049 | 0.4066 | 0.4082 | 0.4099 | 0.4115 | 0.4131 | 0.4147 | 0.4162 | 0.4177 |
| 1.4 | 0.4192 | 0.4207 | 0.4222 | 0.4236 | 0.4251 | 0.4265 | 0.4279 | 0.4292 | 0.4306 | 0.4319 |

(continued)

⊕ Table A.1   (continued)

| Z | α for one-tailed test | | | | | | | | | |
|---|---|---|---|---|---|---|---|---|---|---|
| | 0.00 | 0.01 | 0.02 | 0.03 | 0.04 | 0.05 | 0.06 | 0.07 | 0.08 | 0.09 |
| 1.5 | 0.4332 | 0.4345 | 0.4357 | 0.4370 | 0.4382 | 0.4394 | 0.4406 | 0.4418 | 0.4429 | 0.4441 |
| 1.6 | 0.4452 | 0.4463 | 0.4474 | 0.4484 | 0.4495 | 0.4505 | 0.4515 | 0.4525 | 0.4535 | 0.4545 |
| 1.7 | 0.4554 | 0.4564 | 0.4573 | 0.4582 | 0.4591 | 0.4599 | 0.4608 | 0.4616 | 0.4625 | 0.4633 |
| 1.8 | 0.4641 | 0.4649 | 0.4656 | 0.4664 | 0.4671 | 0.4678 | 0.4686 | 0.4693 | 0.4699 | 0.4706 |
| 1.9 | 0.4713 | 0.4719 | 0.4726 | 0.4732 | 0.4738 | 0.4744 | 0.4750 | 0.4756 | 0.4761 | 0.4767 |
| 2.0 | 0.4772 | 0.4778 | 0.4783 | 0.4788 | 0.4793 | 0.4798 | 0.4803 | 0.4808 | 0.4812 | 0.4817 |
| 2.1 | 0.4821 | 0.4826 | 0.4830 | 0.4834 | 0.4838 | 0.4842 | 0.4846 | 0.4850 | 0.4854 | 0.4857 |
| 2.2 | 0.4861 | 0.4864 | 0.4868 | 0.4871 | 0.4875 | 0.4878 | 0.4881 | 0.4884 | 0.4887 | 0.4890 |
| 2.3 | 0.4893 | 0.4896 | 0.4898 | 0.4901 | 0.4904 | 0.4906 | 0.4909 | 0.4911 | 0.4913 | 0.4916 |
| 2.4 | 0.4918 | 0.4920 | 0.4922 | 0.4925 | 0.4927 | 0.4929 | 0.4931 | 0.4932 | 0.4934 | 0.4936 |
| 2.5 | 0.4938 | 0.4940 | 0.4941 | 0.4943 | 0.4945 | 0.4946 | 0.4948 | 0.4949 | 0.4951 | 0.4952 |
| 2.6 | 0.4953 | 0.4955 | 0.4956 | 0.4957 | 0.4959 | 0.4960 | 0.4961 | 0.4962 | 0.4963 | 0.4964 |
| 2.7 | 0.4965 | 0.4966 | 0.4967 | 0.4968 | 0.4969 | 0.4970 | 0.4971 | 0.4972 | 0.4973 | 0.4974 |
| 2.8 | 0.4974 | 0.4975 | 0.4976 | 0.4977 | 0.4977 | 0.4978 | 0.4979 | 0.4979 | 0.4980 | 0.4981 |
| 2.9 | 0.4981 | 0.4982 | 0.4982 | 0.4983 | 0.4984 | 0.4984 | 0.4985 | 0.4985 | 0.4986 | 0.4986 |
| 3.0 | 0.4987 | 0.4987 | 0.4987 | 0.4988 | 0.4988 | 0.4989 | 0.4989 | 0.4989 | 0.4990 | 0.4990 |

**Table A.2** Critical values of 't'

| df | α for one-tailed test | | | | | | | | | | | | |
|---|---|---|---|---|---|---|---|---|---|---|---|---|---|
| | 0.5 | 0.25 | 0.2 | 0.15 | 0.1 | 0.05 | 0.025 | 0.01 | 0.005 | 0.001 | 0.0005 |
| 1 | 0.000 | 1.000 | 1.376 | 1.963 | 3.078 | 6.314 | 12.710 | 31.820 | 63.660 | 318.310 | 636.620 |
| 2 | 0.000 | 0.816 | 1.061 | 1.386 | 1.886 | 2.920 | 4.303 | 6.965 | 9.925 | 22.327 | 31.599 |
| 3 | 0.000 | 0.765 | 0.978 | 1.250 | 1.638 | 2.353 | 3.182 | 4.541 | 5.841 | 10.215 | 12.924 |
| 4 | 0.000 | 0.741 | 0.941 | 1.190 | 1.533 | 2.132 | 2.776 | 3.747 | 4.604 | 7.173 | 8.610 |
| 5 | 0.000 | 0.727 | 0.920 | 1.156 | 1.476 | 2.015 | 2.571 | 3.365 | 4.032 | 5.893 | 6.869 |
| 6 | 0.000 | 0.718 | 0.906 | 1.134 | 1.440 | 1.943 | 2.447 | 3.143 | 3.707 | 5.208 | 5.959 |
| 7 | 0.000 | 0.711 | 0.896 | 1.119 | 1.415 | 1.895 | 2.365 | 2.998 | 3.499 | 4.785 | 5.408 |
| 8 | 0.000 | 0.706 | 0.889 | 1.108 | 1.397 | 1.860 | 2.306 | 2.896 | 3.355 | 4.501 | 5.041 |
| 9 | 0.000 | 0.703 | 0.883 | 1.100 | 1.383 | 1.833 | 2.262 | 2.821 | 3.250 | 4.297 | 4.781 |
| 10 | 0.000 | 0.700 | 0.879 | 1.093 | 1.372 | 1.812 | 2.228 | 2.764 | 3.169 | 4.144 | 4.587 |
| 11 | 0.000 | 0.697 | 0.876 | 1.088 | 1.363 | 1.796 | 2.201 | 2.718 | 3.106 | 4.025 | 4.437 |
| 12 | 0.000 | 0.695 | 0.873 | 1.083 | 1.356 | 1.782 | 2.179 | 2.681 | 3.055 | 3.930 | 4.318 |
| 13 | 0.000 | 0.694 | 0.870 | 1.079 | 1.350 | 1.771 | 2.160 | 2.650 | 3.012 | 3.852 | 4.221 |
| 14 | 0.000 | 0.692 | 0.868 | 1.076 | 1.345 | 1.761 | 2.145 | 2.624 | 2.977 | 3.787 | 4.140 |
| df | 0.5 | 0.4 | 0.3 | 0.2 | 0.1 | 0.05 | 0.02 | 0.01 | 0.002 | 0.001 |
| | α for two-tailed test | | | | | | | | | | |

(continued)

**⊕ Table A.2** (continued)

| df | α for one-tailed test | | | | | | | | | | |
|---|---|---|---|---|---|---|---|---|---|---|---|
|  | 0.5 | 0.25 | 0.2 | 0.15 | 0.1 | 0.05 | 0.025 | 0.01 | 0.005 | 0.001 | 0.0005 |
| 15 | 0.000 | 0.691 | 0.866 | 1.074 | 1.341 | 1.753 | 2.131 | 2.602 | 2.947 | 3.733 | 4.073 |
| 16 | 0.000 | 0.690 | 0.865 | 1.071 | 1.337 | 1.746 | 2.120 | 2.583 | 2.921 | 3.686 | 4.015 |
| 17 | 0.000 | 0.689 | 0.863 | 1.069 | 1.333 | 1.740 | 2.110 | 2.567 | 2.898 | 3.646 | 3.965 |
| 18 | 0.000 | 0.688 | 0.862 | 1.067 | 1.330 | 1.734 | 2.101 | 2.552 | 2.878 | 3.610 | 3.922 |
| 19 | 0.000 | 0.688 | 0.861 | 1.066 | 1.328 | 1.729 | 2.093 | 2.539 | 2.861 | 3.579 | 3.883 |
| 20 | 0.000 | 0.687 | 0.860 | 1.064 | 1.325 | 1.725 | 2.086 | 2.528 | 2.845 | 3.552 | 3.850 |
| 21 | 0.000 | 0.686 | 0.859 | 1.063 | 1.323 | 1.721 | 2.080 | 2.518 | 2.831 | 3.527 | 3.819 |
| 22 | 0.000 | 0.686 | 0.858 | 1.061 | 1.321 | 1.717 | 2.074 | 2.508 | 2.819 | 3.505 | 3.792 |
| 23 | 0.000 | 0.685 | 0.858 | 1.060 | 1.319 | 1.714 | 2.069 | 2.500 | 2.807 | 3.485 | 3.768 |
| 24 | 0.000 | 0.685 | 0.857 | 1.059 | 1.318 | 1.711 | 2.064 | 2.492 | 2.797 | 3.467 | 3.745 |
| 25 | 0.000 | 0.684 | 0.856 | 1.058 | 1.316 | 1.708 | 2.060 | 2.485 | 2.787 | 3.450 | 3.725 |
| 26 | 0.000 | 0.684 | 0.856 | 1.058 | 1.315 | 1.706 | 2.056 | 2.479 | 2.779 | 3.435 | 3.707 |
| 27 | 0.000 | 0.684 | 0.855 | 1.057 | 1.314 | 1.703 | 2.052 | 2.473 | 2.771 | 3.421 | 3.690 |
| 28 | 0.000 | 0.683 | 0.855 | 1.056 | 1.313 | 1.701 | 2.048 | 2.467 | 2.763 | 3.408 | 3.674 |
| 29 | 0.000 | 0.683 | 0.854 | 1.055 | 1.311 | 1.699 | 2.045 | 2.462 | 2.756 | 3.396 | 3.659 |
| 30 | 0.000 | 0.683 | 0.854 | 1.055 | 1.310 | 1.697 | 2.042 | 2.457 | 2.750 | 3.385 | 3.646 |
| 40 | 0.000 | 0.681 | 0.851 | 1.050 | 1.303 | 1.684 | 2.021 | 2.423 | 2.704 | 3.307 | 3.551 |
| df | 1 | 0.5 | 0.4 | 0.3 | 0.2 | 0.1 | 0.05 | 0.02 | 0.01 | 0.002 | 0.001 |
|  | α for two-tailed test | | | | | | | | | | |

● **Table A.2** (continued)

| df | α for one-tailed test | | | | | | | | | | |
|---|---|---|---|---|---|---|---|---|---|---|---|
| | 0.5 | 0.25 | 0.2 | 0.15 | 0.1 | 0.05 | 0.025 | 0.01 | 0.005 | 0.001 | 0.0005 |
| 60 | 0.000 | 0.679 | 0.848 | 1.045 | 1.296 | 1.671 | 2.000 | 2.390 | 2.660 | 3.232 | 3.460 |
| 80 | 0.000 | 0.678 | 0.846 | 1.043 | 1.292 | 1.664 | 1.990 | 2.374 | 2.639 | 3.195 | 3.416 |
| 100 | 0.000 | 0.677 | 0.845 | 1.042 | 1.290 | 1.660 | 1.984 | 2.364 | 2.626 | 3.174 | 3.390 |
| 1000 | 0.000 | 0.675 | 0.842 | 1.037 | 1.282 | 1.646 | 1.962 | 2.330 | 2.581 | 3.098 | 3.300 |
| ∞ | 0.000 | 0.674 | 0.842 | 1.036 | 1.282 | 1.645 | 1.960 | 2.326 | 2.576 | 3.090 | 3.291 |
| df | 1 | 0.5 | 0.4 | 0.3 | 0.2 | 0.1 | 0.05 | 0.02 | 0.01 | 0.002 | 0.001 |
| | α for two-tailed test | | | | | | | | | | |

Printed in the United States
by Baker & Taylor Publisher Services